James Edwards Smith

How to See with the Microscope

James Edwards Smith

How to See with the Microscope

ISBN/EAN: 9783744692304

Printed in Europe, USA, Canada, Australia, Japan

Cover: Foto ©berggeist007 / pixelio.de

More available books at **www.hansebooks.com**

HOW TO SEE

WITH

THE MICROSCOPE:

BEING

USEFUL HINTS CONNECTED WITH THE SELECTION AND USE
OF THE INSTRUMENT; ALSO SOME DISCUSSION OF THE
CLAIMS AND CAPACITY OF THE MODERN HIGH-
ANGLED OBJECTIVES, AS COMPARED WITH
THOSE OF MEDIUM APERTURE; WITH
INSTRUCTIONS AS TO THE SELEC-
TION AND USE OF AMERI-
CAN OBJECT - GLASSES
OF WIDE APER-
TURES.

BY

J. EDWARDS SMITH, M. D.,

PROFESSOR OF HISTOLOGY AND MICROSCOPY IN THE CLEVELAND (O.), HOM-
ŒOPATHIC HOSPITAL COLLEGE; CORRESPONDING MEMBER OF THE SAN
FRANCISCO, THE DUNKIRK, AND ILLINOIS STATE MICROSCOPICAL
SOCIETIES; MEMBER OF THE AMERICAN ASSOCIATION
FOR THE ADVANCEMENT OF SCIENCE, ETC.

ILLUSTRATED.

CHICAGO:
DUNCAN BROTHERS,
1880.

DEDICATION:

TO

THE HON. P. H. WATSON,

WHOSE FRIENDSHIP HAS ENCOURAGED AND WHOSE PARTICIPATION HAS

GIVEN DOUBLE PLEASURE TO THE INVESTIGATIONS RECORDED

IN THESE PAGES; THE TRUE FRIEND, AND THE LIB-

ERAL PATHON OF SCIENTIFIC EXPERI-

MENT, THIS LITTLE BOOK

IS DEDICATED

BY THE

AUTHOR.

PREFACE.

INTRODUCTORY AND APOLOGETICAL.

In the spring of 1874 I received from R. B. Tolles, Esq., (the well-known optician of Boston, Mass.,) a one-sixth immersion object glass, which he requested me to study carefully, and to report the results to him.

The angular aperture of this new one-sixth was said to be 180°, and the objective was one of the first " duplex " or four-system glasses devised by Mr. Tolles, and destined, in a few short months, to create quite a stir in microscope circles.

Having, at that date, leisure at my command, and being very much interested in the performance of object glasses, I was very glad to give this objective careful and prolonged attention. The result was, that after thirty days' experience I had arrived at a settled conviction that Mr. Tolles had made a decided advance in the construction and performance of microscope object glasses.

Believing this experience of mine to be of value to microscopists, I wrote a short account of the performance of this one-sixth, which was published in the *Cincinnati Medical News.*

Shortly afterwards Mr. Tolles kindly sent me another of the " duplex " or four-system objectives, this time a one-tenth. This glass also occupied my leisure time for a considerable interval.

The performance of the one-tenth fully sustained the high appreciation I had formed of the performance of the duplex glasses. In some respects the work of the one-tenth excelled that of the one-sixth.

The article published in the *Cincinnati Medical News* was therefore supplemented by several other communications written by me, giving my further experience with the duplex objectives.

Among the conclusions I had arrived at from the study of the new four-system glasses were to be found embodied ideas radically opposed to popular dogmas which were at that date fully received and accepted by microscopists generally as settled matters of fact. For instance, it was claimed for the duplex one-sixth and one-tenth that either of these objectives would not only do the work of, but excel the performance (under amplifications from 275 to 4,000 diameters) of any one-fiftieth extant. Furthermore, it was claimed that the central light work of the new wide apertured duplex objectives would surpass that of the low angles.

Referring to the resolution of the severest test objects, it was the express purpose of the said contributions to the *Cincinnati Medical News* to claim *and insist on, as a stubborn fact*, that the work of the duplex one-sixth or one-tenth excelled that of any glass extant, " be it a one-fifth or a one-fiftieth."

These, as well as other statements so thoroughly heterodox to established belief, met, as a matter of course, with earnest opposition from my readers. A large correspondence ensued, while many microscopists visited me for the purpose of witnessing, by way of ocular demonstrations, the performance of the duplex g.asses.

Meanwhile, in London, England, a lively controversy, known in microscope circles as "the war of the apertures," appeared in the columns of the *London Monthly Journal of Microscopy*, in which controversy the positions assumed by Mr. Tolles were assailed by Mr. F. A. Wenham. The issues involved, however, appertained only to *optical possibilities*, the performance of the duplex objectives being an entirely outside matter. During the London battle Mr. Tolles was ably assisted by Dr. J. J. Woodward, of Washington, D. C., and by Prof. Keith of Georgetown. Suffice it here to say, that from the standpoint of optical science Mr. Tolles maintained his positions, a fact, I believe, now generally admitted.

And now, after a period of nearly six years, it is to me a matter of pride to here record that of the several "heterodox" positions I had assumed in public print, there has occurred, neither in the interim, or at this present writing, no occasion to retract one word of what was claimed by me in my former contributions to the *Cincinnati Medical News*. The duplex objectives have steadily, and perforce of intrinsic worth, forced their way into general use, while the leading opticians are exerting themselves to still further improve their performance.

Having, as I have above briefly set forth, had a very large experience with the four-system objectives, it was suggested to the author that a small work, giving in detail the manipulations of the new objectives, with, perhaps, some other items resulting from an experience of over fifteen years as a microscopist, would not only be acceptable, but would fill a place at present unoccupied; and it is in response to this suggestion this little book is presented, with the hope that it may prove of some service to some of my brother microscopists.

The book is entirely innocent of literary pretensions. It has been my aim to express myself in the simplest possible manner. The diction is at times redundant, gossipy and commonplace, the dominant idea throughout being to hold a good natured chat with my readers.

During the past eight years the author has received thousands of letters asking for information in the matter of microscope object glasses, and especially has his experience been called on as to methods of *manipulation* involved in the use of wide apertured objectives. In the following pages he has endeavored to supply this information, introducing, it is believed, the first attempt to teach the difficult art of collar adjustment, as will be seen in the graduated series of lessons, which, imperfect as they are, it is hoped will nevertheless be of value to those commencing to use the four-system objectives.

And now, by way of apology, it remains to state that circumstances entirely unlooked for, and entirely beyond the author's control, have caused a delay in the publication, and many have suffered disappointment therefrom. This delay, however, has furnished me with the opportunity to revise the MS., rendering it, I trust, more acceptable. It has also given me the opportunity to introduce new matter not originally contemplated.

During the period of delay above mentioned the writer was in receipt of hundreds of letters of enquiry from as many kind friends. It was impossible to reply to all of these. He, however, improves the present as a fitting occasion to return his grateful thanks to those of his correspondents for the tangible evidences of confidence and esteem thus manifested.

And to the Messrs. Duncan Bros., too, is he under obligations for the hearty manner in which they took the little *brochure* under their wing, pushing the same through the press with their accustomed energy.

323 Euclid Avenue, Cleveland, O., September 1880.

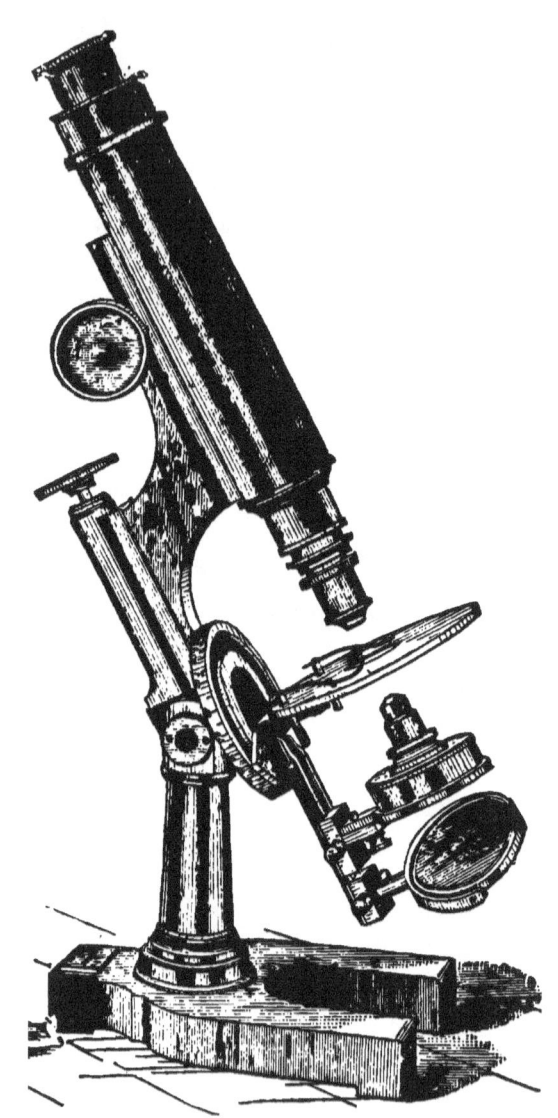

ACME STAND.

CONTENTS.

——:o:——

	PAGE.
Introduction and Apologetical	5

CHAPTER I.

Something about American Stands, etc.	17
Zentmayer's American Centennial Stands. . . .	34
The American Histological Stand	39
Tolles' large " B B " Microscope	47
Tolles' Largest " A " Microscope	48
Tolles' Student's Microscope	49
The Professional Microscope	53
Large Student's Microscope	46
Family Microscope	57
Bulloch's First-Class Microscope	59
Small best Stand " A B "	67
Mr. Bulloch's " D " Stand	67
Mr. Bulloch's New Biological Stand	68
R. &. J. Beck's Microscope	71
Beck's Popular Microscope, Minocular or Binocular .	72
Beck's Economic Microscope	72
Beck's New Histological Dissecting Microscope . .	79
The New National Microscope	80
The New Acme Stand	84

13

CHAPTER II.

What is Augular Aperture 93
Angular Aperture 93
How Shall we Measure Angular Aperture . . . 95
Object Glasses . , ... 97
Something Further about Objectives 108
Balsam Apertures . . . . · 123
Flatness of Field . , ... 126
Mounting of Objectives 129
Nomenclature of Objectives . . . . . . 130 ,
Table of the Magnifying Power of Single Convex Lenses 134

CHAPTER III.

Objectives Continued . · 137
Objectives of Lower Balsam Angle 144
Adjustable Objectives . · . , ... 148
Eye-Pieces . · 156

CHAPTER IV.

Manipulations — Wenham's Reflex Illuminator . . 157
The Woodward Illuminator . , ... 163
Tolles' Traverse Lens 179

CHAPTER V.

Illumination 183
Sunlight 186
Artificial Light 187

CHAPTER VI.

Choice of Objectives for Regular Work . . . 202
Selection of Covering Glass 213
Bull's Eye Condenser 225
Working with Low Powers 225
The Spencer one-inch of 50° — Broad Guage Objectives,
    etc. 230

CHAPTER VII.

Work with the Higher " Powers " 235
Position of Observer 246
Mean of ten Measurement of Moller Test Plates . . 253
Position of Observer 289
Work over dry Mounts with high Aperture Objectives . 298
Oil Immersion Object Glasses 309
New Oil Immersion Objectives 310

CHAPTER VIII.

A Word or two on Volumetric Analysis . . . 318
Apparatus necessary 320
Analysis of Urea, etc. 325
Analysis of Sugar, etc. 332
Preparation of Urinary Constituents 339

APPENDIX.

Names and Address of Dealers in Microscopes, Object-
  ives, etc., Alphabetically Arranged 341
The Investigator Microscope 345
International Microscope Stand, ... 348
Prices of Accessories 352

SUPPLEMENT.

Contributions to the Cincinnati Medical News . . 358
High Angles . . . , ... 359
Angular Aperture and Central Illumination . . 361
On the Performance of Objections 368
Angular Aperture once More 370
The use and abuse of the Microscope 375
A Chapter on Elementary Physics . , . . . 393
Choice of Objectives . , , ... 397
Microscopical Examinations of Oleomargarine . . . 400

HOW TO SEE WITH THE MICROSCOPE.

CHAPTER I.

SOMETHING ABOUT AMERICAN STANDS, ETC.

The choice of a stand is a matter of interest to every working microscopist, and to most of us it has been a source of much annoyance and needless expense. It may be safely affirmed that most workers have wasted more money in previous purchases of unsatisfactory stands than would suffice to pay for the one at present in use; very many indeed would feel grateful to be let off with this record. I therefore propose to assist the *novice* in the important matter of the selection of a stand, and shall, with this end purely in view, give him fearlessly what, in my opinion, will be sound advice.

And be it known again that there will be some advertising done in this department. As I have before intimated, it is quite impossible to say what I shall have to say, of real value to the reader, without giving prominence to one or more of the several makers. The responsibility is mine, and I accept it. It has been my province, in times past, to select many stands for my friends and acquaintances; those who have visited me with the special object to ascertain my methods of using and working objectives have, in many instances, fol-

lowed my recommendations as to stands, and in every
case, I believe, have expressed to me their satisfaction.
The author has not in the past stood so much alone as
to his ideas of a stand as was the case with his opinion
of objectives.

Now, in what follows, it is to be assumed that the
reader desires *one* stand, and one only, and that he
wishes to invest his money to the best advantage, *i. e.*,
make the same go as far as possible; that he desires a
really good and reliable instrument, and one that will
last for a lifetime.

During the past three or four years, the microscope
stand has been greatly improved, both in Europe and
America, and as a rule, really serviceable instruments
can now be obtained either of American or London
workmanship, and at a moderate cost.   It is to be regretted that our German friends have not followed suit,
but, on the contrary, have been content with the old
form of stands which were patent ten years ago.

In the late improvements by the American and London makers, it is first noticeable that the weight of the
stand has been considerably reduced; the old idea, that
to secure sufficient solidity it was necessary to employ
stands weighing from twenty to forty pounds, being
now practically abandoned.

Again, it was formerly considered a *sine qua non* that
the microscope stage be thick enough, heavy enough,
and solid enough to bear the whacks from a sledge-
hammer.

It must be here kept in mind that years ago there

were no such things as wide-angled objectives, and, as a matter of course, the work was principally done with central or centrally disposed light. To this the old and heavy stage offered no objection, for it was quite possible, with the aid of achromatic condensers, prisms, etc., to work with all the obliquity the objective would respond to, using a stage two or three inches in thickness. Luckily, there were those who would at times "fight objectives," play with diatoms, etc., and in response to their demands the optician increased his angles and working force of the object-glasses. To meet this in turn called for the construction of condensers of greater angle; until finally it occurred that the aperture of the objective had arrived at proportions to which the condensers did not satisfactorily respond. Something had to be done, and something was done, for necessity is the "mother of invention."

The simplest way is the best, and this was to reduce the thickness of the stage. The "fighter" of objec- tives had discovered the fact that a stage, one-fifth of an inch in thickness, was solid enough for any and all of the delicate work required by the microscopist, while at the same time he derived a great advantage in thus pro- viding play for the aperture of his objectives.

The writer remembers with pride that he "took a hand" at this; he remembers, too, the unalloyed satis- faction he experienced in seeing two heavy, lumbering, and expensive stages, alone costing several hundred dol- lars, removed from imported stands, and their place substituted by plain, thin plates made by the local watchmaker; he recollects, also, the exclamation of one of his friends, after looking at an object as displayed by the improvised stage: "Well, I declare! This instrument *was made* for any thing in creation but to see through."

Hence it was, responding to the increasing call, our American and London makers decreased the thickness of their stages. While it is yet true that many of the present stages are unnecessarily thick, the reduction is still palpably manifest.

It may be remarked, then, that the real improvement of late years in the construction of American and London stands may, as a rule, be manifested in these two items, viz., reduction of weight and thickness of stage.

We have learned something from the Germans, too, within the past few years. It has been well known that they favored the vertical stand with its short tube. Our experience has taught us that both of these have their advantages, and our later instruments are so contrived as to be used with short, and also with standard tube, and in a vertical or inclined position.

I desire in this place to record the fact that the stands made in the United States are not excelled in any quality or condition going to make a number one, reliable instrument. The stands produced by our home makers are quite equal in every respect to those of any other countries, while their cost is not one whit higher.

Those contemplating the purchase of a stand will, as a matter of course, consult their individual taste and inclinations to a considerable extent: thus, A may select a large and heavy stand, while B would prefer a smaller and lighter one. The party, too, will naturally take into consideration the particular field of work he may have in contemplation. It is quite possible to give the latter consideration too much weight. It will be found, as a rule, better policy in the long-run to purchase a stand capable of doing almost any work, and thus avoid the possibility of being compelled to sell at some future date, at a heavy discount, and purchase another and more capable stand. Fortunate it is that one can now purchase, at moderate figures, reliable and well-made stands, suitable for almost any purpose of the microscopist.

In the reference that has been made to American stands, it is proper to state that the Messrs. Beck, of London, have a regular agency here for the sale of their wares, and that this agency is in charge of an American gentleman. There can be no good reason to regard them as other than home folks. At all events, the author is of this opinion, and will act accordingly.

Among the essentials that deserve attention in the selection of a reliable stand may be mentioned:

FIRST. See that the stand is well balanced in *every* position that it manifests no disposition to topple either one way or the other; that it stands tolerably firm (for its weight) on its legs.

SECOND. If it has coarse adjustment by rack and pinion, see that the movement is as smooth as oil; reverse the milled head between the thumb and finger promptly, and notice if there be any lost motion; try this test all the way of the "run" of the rack, and if there be "lost motion" detected discard the stand.

THIRD. Examine *carefully* the bearings on which the body slides; these should be broad, and the fitting of the body to the limb accurate; test by placing a small object on the stage — a circular diatom will be the thing; now examine with a half-inch glass; rack the tube up and down a little, and see if the object keeps its centre; seize the body-tube near the eye-piece, and twist it a bit from right to left, and *vice versa*, noticing whether the object "travels" or not; repeat this experiment by punching holes in a card-box, say three-quarters of an inch thick, thus forming a supplemental stage; focus again and try the twist once more. The instrument that will stand this test is all right as to its bearings.

FOURTH. Place the diatom on the stage, under the half-inch, as before, and if the instrument has fine adjustment by nose-piece, test by moving the fine wheel quickly either way, and see if the object "travels." If satisfactory with the half-inch, try an eighth, or tenth; test also by taking hold of the objective as one would in the adjustment for cover, by giving it a little twist right and left, and see if the object changes its position. Finally, examine the run of the fine adjustment; see if it is quick and sensitive to the touch, without "jump;" if satisfactory so far, so good.

FIFTH. Should the stand have concentric stage, test the accuracy of its fittings with diatom and half-inch glass. If the stand has suffered transportation, or has in any way been submitted to fatigue, the probability will be that the stage has become somewhat decentred, and any effort of your own to re-adjust will be likely to make things worse. The very best stages will not remain centred long if roughly used. Still, with your half-inch and the diatom you can manage to get some idea, which will be better than none at all. Centre your diatom, and focus; now seize opposite edges of the stage with both hands, hold it firmly, giving it "jerks" either way, right and left, but not sufficiently to absolutely move the stage on its centre. If you succeed in moving your object, with a corresponding movement of the stage, then the latter is not accurately fitted, and cannot be accurately centred. This "slip" can be detected with the hands alone.

SIXTH. It the stand be furnished with mirror fitted to radial arm, see that all these fittings are strong and likely to last. This portion of the stand meets with more fatigue than any other. See that the friction surfaces are *not* both the same metal; this will apply, too, to any part of the stand. Notice particularly whether the universal motions of the mirror are properly constructed, and likely to last for years; this is a most important point.

FINALLY. It will be well to see that the joint for inclining the instrument at various angles is strong and well made, and that it have compensation for wear. Notice the general "get up" — the general finish of its various parts. And now, having received a lesson on *the use of diatoms*, you may put the little fellow away until again wanted.

Aside from these matters involving sound mechanism, there remain other points connected with the choice of a stand, which may, to some extent, be regarded as a matter of election. For instance, some of the most costly stands are furnished with mechanical stages whereby motions are given to the object-carrier by various milled heads. I have used these stages in times past, and have to record my disapprobation of them, and for the following reasons:

First. They are an impediment to quick work. It is much quicker to run from one end of an object to another by one single movement given by hand than to wait the slow motions of the screws. There are, however, some advantages arising at times from the use of the mechanical stage, e. g., in adjusting the image of an object to the eye-piece micrometer, etc. Nevertheless these slight conveniences are sadly outweighed by the positive objection to their use.

Second. A mechanical stage, to be good for any thing must be nicely made; hence they are costly, and further, seldom keep in order for a great length of time, *however well made.*

Third. As generally modelled, they increase the thickness of the stage, and the screws are *always* more or less in the way.

Fourth. Those who rely on their mechanical assistance seldom arrive at that delicate *finger* manipulation so necessary to be acquired by the observer. Any one of the objections above named ought, in my opinion, to be sufficient and determinate.

Many of the first-class stands (so called) are fitted with sub-stages, provided with rack and pinion, and centering screws. In the latter models these accessories can be completely removed, leaving the entire instrument below the stage clear and unobstructed. Beyond the cost which these appliances incur, I see no particular objection to their presence; but the novice is informed that they are by no means a *necessary* adjunct, and that their duties can be performed by simple and less costly contrivances.

Stands have lately been introduced with short tubes, but capable, by means of an interior draw, of being drawn out to the standard length of ten inches. This is an improvement on the short tube of the German school, and one of real value. The daily worker will find the short tube a great convenience when working over wet preparations, or dealing with reagents. Under these circumstances, to work quickly one must needs keep the stage horizontal and the tube vertical. It is true that we can compensate, in a measure, in using the standard lengths by working with a lower table; but in this case the observer cannot have his instrument so well under control, and he is further compelled to resort to some out-of-the-way contrivance by way of getting rests for the elbows.

In the selection of a stand, I advise strongly that it be fitted with a concentric rotating stage; and should the *res angusta domi* present no obstacle, let the stage be a good one, divided to degrees, capable of an entire revolution, and furnished with an object-carrier. The

advantages of the revolving stage are manifold, while
to the advanced worker, it is a *sine qua non*. In the
examination of difficult and resisting structures the observer needs at times to employ almost every variety of
illumination, as well as to view the object under various
aspects. In this department of work the rotating stage
cannot be dispensed with.

Where money *is* an object, let the would-be purchaser
proceed with due care, giving this matter the closest
attention. By observing the following instructions he
can provide for a rotating stage capable of responding
to almost any call, and at an outlay not to exceed five
dollars. The writer has two stands. One is a large,
heavy, first-class instrument, and is, of course, furnished
with adjusting concentric stages, divided to half-degrees;
the other is one of the smallest stands made, and is
fitted with short tube (which can be drawn out to
standard length), and a plain, revolving stage, which
cost *two dollars*. *Ninety-five* per cent. of all his work
is done on the little stand; the work allotted to the
larger instrument being the measurement of the angles
of crystals, recording of objects by the Maltwood finder,
and the measurements of the angular aperture of objectives. This much for the solace of those who are
bothered with the hard times.

In making the selection with the view in hand of
providing a cheap rotating stage, the first important
point is to see that the instrument has sufficient stage
room. The stage should measure one and seven-eighth
inches from the centre of well-hole to the nearest por-

tion of the limb touching the stage; and any instrument having less distance than this will not answer for the purpose, while a half-inch more room would be very desirable. The object of all this is to get room enough, so that we can employ a circular stage large enough to work an object-carrier.

. Next, see that there are no sub-stage fittings, which, by their particular method of attachment, will interfere with the stage about to be described. Look the ground over carefully, and, accepting that the road is all clear, we proceed to describe the stage which the author has had in daily use for years, and one that has, to a considerable extent, been copied by his friends. Any watchmaker or machinist can do the work — in fact, one of my friends made one for himself.

Provide a sheet of well-hammered brass, heavy enough, so that when planed or turned down the stage shall be one-sixteenth of an inch in thickness, with both faces truly parallel. Cut the circle which is to form your stage as large as your instrument will permit, and in accordance with the above directions.

Cut the well-hole one-sixteenth of an inch larger than the well-hole of your stage; make a collar, or short tube, out of the same material used for the stage; turn the outside to proper dimensions, so as to fit the well-hole of the new stage, the upper edges of both being "flush," and solder in position.

Next: Turn accurately the under and projecting part of the short collar, or tube, so that it will *exactly* fit the well-hole of your main stage; place it thereon, and cut off any portion of the collar that may project beneath said stage.

In the stage thus far towards completion, it so be that the collar projects one-sixteenth of an inch, this will be found ample for its support; thus you will be enabled with some stands to steer clear of sub-stage appliances, etc.

All that remains to be done is to fit your new stage with plain spring-clips, which can be done in a few moments out of a piece of watch-spring; or, if there is room enough, you can provide an object-carrier, made on Mr. Zentmayer's principle; all of which is plain work, and easily accomplished by any machinist of tolerable skill.

Keep in mind that this stage adds somewhat to the thickness, and govern yourself accordingly.

" But," says one, " I have no assurance that this stage will rotate in the optical axis." I grant it, with the remark that if it did so rotate with one objective, it would be pretty sure to fail with another. The compensation must be supplied by finger manipulations, easily acquired, and as easily practiced. As I have previously hinted, the best rotating stages remain concentric but for a short time, especially if much used; while to the real worker, the very bother of adjusting the most expensive stage extant would be an intolerable annoyance, and a willful waste of time.

A grand good thing about this improvised stage is, that it can be placed in position or removed therefrom in a moment's time. This, to the author, is a real boon, for he has often to remove the supplemental stage, and to work for hours with the mounts placed on the main plate, using not even the spring-clip or any other attachment, the microscope being the while in the vertical position.

Says another: Why not have the maker furnish some such stage at the date of purchase? He can do these things better than any one else." I respond, Yea, verily.

Thousands of times the question has been asked, "Which do you prefer—the binocular or the monocular?" and as it is more than probable that this question will arise in the minds of some of those who read this book, perhaps a word or two on the subject may not be amiss.

Mr. Henry Crouch, F. R. M. S., a well known maker of microscopes, visited this country during the Centennial Exhibition, and on his return complained bitterly of "an eminent German microscopist, who assisted in examing the microscopes on exhibition at Philadelphia, and who from the first loudly proclaimed the uselessness of binoculars, . . . but whom he afterwards found out had never used one." The author is pretty much in the same boat with the eminent German.

Since the introduction of the binocular, the writer has made several downright square and honest attempts to use the binocular long enough to be able to express an opinion worth consideration, but in each and every case the double barrelled machine proved too much for his patience. With the low powers the instrument is capable of giving very pretty shows, particularly where a certain stereoscopic effect is supposed to add force. Hence the binocular is eminently fitted for such displays, and is eminently adapted, too, for the entertainment of those of our lady friends who visit the soirees of microscopical societies, without feeling specially interested in microscopy as a science.

If so be that there can be structure displayed by the binocular that cannot be seen equally well by the monocular, be it so, and the author will joyfully add his testimony to the same, *after having arrived at the fact.*

Many of my friends have binocular instruments, but I can not recall the name of any person who uses one; that is, when I am present, for inevitably the first thing done by the owner is to displace the prism, and to *use* the stand as a monocular. I have never found a solitary exception to this rule.

But some one may say: "What possible objection can there be to purchasing a binocular, since, as you admit, the instrument can be changed instantly to a monocular, and the purchaser has his choice. He certainly has all that you have got, and perhaps (allowing force to the opinion of others) more too."

Well, that looks lucid enough; and in truth that argument has sold many a binocular. Nevertheless, "all is not gold that glitters." Let us take a look from another stand point, thus:

First. The binocular involves greater weight; as a rule, they are unwieldly, lumbering things, destitute of grace, symmetry, or aught that goes to make a clean, well proportioned stand.

Second. The binocular instrument is much more complicated, and hence more expensive, the purchaser being required to pay, without getting in return value received.

Third. This form does not admit of the instrument being used with short, as well as standard tube, and to the " worker " this item has particular force.

Fourth. The extra expense attending the purchase of a binocular can be better applied in other directions. Those who value dollars and cents will find force in this objection.

Fifth. I oppose the binocular, because the monocular is *good enough,* and because the real work of the microscope has been and will continue to be done with the monocular.

Sixth. The binocular is to a certain extent impracticable, because the two eyes of the observer are not alike. There are exceptions, but not to an extent sufficient to invalidate the rule.

Seventh. I oppose the usual form of binocular instruments, believing that the binocular eyepiece invented and made by Mr. R. B. Tolles is a preferable way of obtaining binocular vision.

Having thus presented my objections, let the reader elect for himself, with this assurance on my part, viz., better get a good binocular than a poor monocular.

All stands should be furnished with plane and concave mirrors. If the mirrors are attached to radial arm, hinged at a greater or less distance from the under surface of the stage, then it is apparent that when the

radial arm is placed in position for oblique light the
mirror will be nearer the well hole than when the arm
is centrally disposed; therefore, the mirror should be
arranged to slide on the arm, so that the proper compensation may be effected.

The mirror should, of course, be well mounted in its
own semicircular arm, so that universal motions may be
obtained, and the semicircle should be closely attached
to the slide moving on the radial arm; that is to say,
there should be no intermediate, short, and jointed arms
or elbows such as are found on many so-called first
class instruments, and are a most intolerable nuisance.
It is quite possible at times to remove these intermediate arms, and attach the mirrors directly to the radial
arm, and at a trifling expense, thus transforming a
faulty stand into a really serviceable one.

Examine the concave mirror carefully, as to its focal
length. To do this, place a piece of white letter paper
on the stage, and using a common candle reflect the
light on the paper; now move the mirror up and down
on the radial arm, and see if you can get a tolerably
well defined image of the flame. It may occur the
radial arm is too short for the focal length of the
mirror, and that it would be impossible to lengthen it
sufficiently; this being the case, reject the stand.

One would naturally suppose that an item like the
last would surely be attended to on the part of the
maker; nevertheless, many stands are made and sold
with glaring defects of this character.

At the risk of being tiresome, let me especially insist

on the absolute importance of looking carefully to the entire apparatus connected with the hanging of the mirrors. This part of the mechanism is called to endure more fatigue, and is oftener thus called on than any other portion of the stand. For instance: In the act of observing, one often has occasion to seize the mirror, and to move it and the radial arm by a single impulse, simultaneously as it were; and here let me say that it will be well for the observer to acquire this habit; nevertheless, it is terribly straining on the joints thus called on, and they should be strong enough to endure the fatigue.

See also that all the eye pieces slip in and out of the tube easily; this too without decentring the object in the field. Any difficulty in putting eye pieces in place or removing therefrom is a first class botheration.

We now proceed to the description of the stands made and sold by American makers. In placing Mr. Joseph Zentmayer at the head of this list, the author feels assured that he does no violence to the feelings of others. Mr. Zentmayer was one of our earliest and most energetic makers, and his work has, as a rule, proved honest and reliable. In his expenditure of capital, his facilities for execution, or his pride in presenting first-class work, he stands second to none, as his competitors will generally attest.

Mr. Zentmayer's largest and most costly stand has, I may say, but just been placed on the market. It was especially designed for exhibition at the Centennial Exposition. The description which follows is to a con-

siderable extent the same as that printed in Mr. Zentmayer's catalogue — some portions being suppressed,
and the wording occasionally changed at the election of
the author.*

ZENTMAYER'S AMERICAN CENTENNIAL STAND.

Constructed especially for the Centennial Exhibition.
It is mounted on a tripod, with revolving graduated
platform ; the bar and trunnions are in one piece, and
swing between two pillars for inclining the instrument
to any angle. The coarse adjustment is accomplished
by rack and pinion. Thus far it is similar to the
"Grand American Stand" by the same maker.

The swinging sub-stage, which carries the condenser
or other illuminating apparatus, including the mirror,
swings around a pivot, the axis of which passes through
the object observed, so that this object is in every position in the *focus of illumination*. The stage may be
detached with facility, and replaced by one constructed
for oblique illumination; the swinging illuminator
may then (*i. e.*, with the last-named stage) be used for
illumination from above.

The sub-stage is provided with a graduated circle for
indicating the degree of obliquity.

An object placed on the stage being in a plane with
the axis of the trunnions, it is obvious that if the instrument is placed in a horizontal position, the object

---

* In the several descriptions of stands, that of the respective makers will
be given as far as possible. The author will, however, at his election, sup-
press certain portions or change the diction as to him may appear desirable.

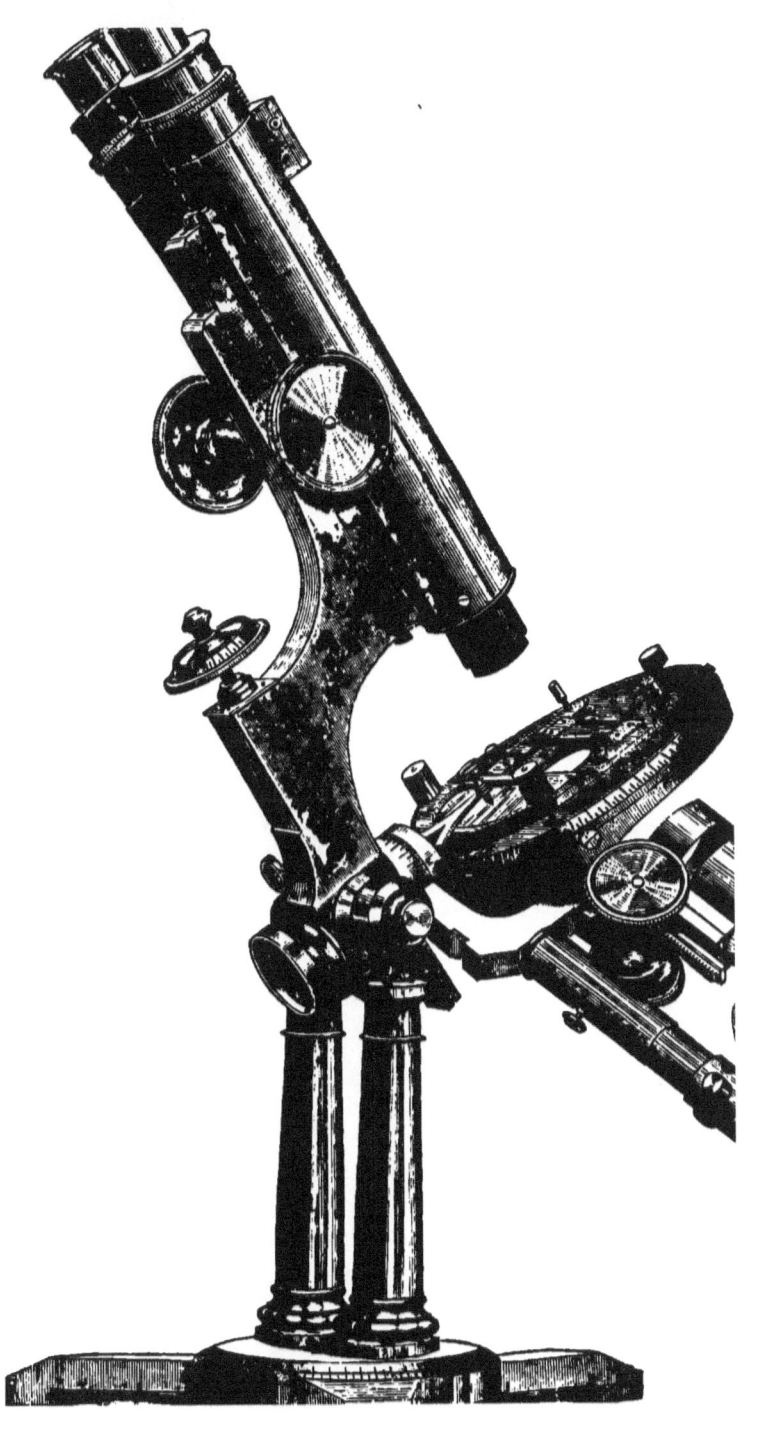

is in the axis of revolution of the graduated platform, and the angular aperture of an objective focused on this object can easily be measured. It is equally obvious that in this position the object is in the centre of all the revolving parts of the instrument, to wit, the revolving stage, swinging sub-stage, and the platform.

The principal stage is similar to the circular one previously used on the " Grand American;" it is provided with adjusting screws for accurate centring, and revolves in a large outside ring, giving facilities for oblique illumination up to 70 degrees from axis (140 degrees aperture), while the graduations serve as a goniometer for the measurement of crystals, etc.

The sub-stage is divided into *two* cylindrical receivers, to facilitate the adaptation of several accessories at one and the same time; the lower cylinder can be moved up and down or entirely removed.

The fine adjustment (in all other instruments of the Jackson model being in front of the body) is removed to the more stable part of the stand. The bar is provided with two slides, one for the rack and pinion movement, and close to it another one of nearly the same length for the fine adjustment, moved by a lever concealed in the bent arm of the bar, and acted on by a micrometer screw. Thus the body is not touched when using the fine adjustment, and the relative distances of objective, binocular prism and eye-piece remain unchanged.

The smaller stage of the American Centennial Stand is also provided with screws for accurate centring; this stage is three inches in diameter and extremely thin, allowing, in connection with the swinging sub-stage and mirror, not only the greatest obliquity of illumination, but the mirror and achromatic condenser will rise above the stage when required, as in the case of sunlight illumination, that of opaque objects, etc.

The diameter of the sub-stage is the same as that of the " Grand American;" the accessories of that stand are therefore interchangeable.

As to the general character of Mr. Zentmayer's work, the author can affirm with confidence that it is not excelled in any particular. The stand just described is beautiful in design, is nicely proportioned, and in every repect reliable and durable. It will stand all of the tests named in the preceding pages. Those wishing a first-class stand cannot fail to be satisfied with the Centennial.

The swinging sub-stage carrying the mirror, etc., is a most valuable improvement, and one that the observer can hardly afford to be without; the mechanism, too, by which this end is accomplished is of the strongest and most workman-like order.

Attention also is invited to the method of attachment of the stage. The latter is solidly held in position, or can in a moment be detached, and another stage substituted.

It has occurred to the writer that the " principal stage " mentioned might very well be dispensed with the smaller stage being quite sufficient. Possibly a large and plain stage might at times be found a convenience, this could easily be substituted for the more expensive one furnished with the instrument.

As to the smaller stage referred to, the author can "speak by the card." He had used one of Mr. Zentmayer's army hospital stands for years, and the instrument gave him good satisfaction. Two years ago, however, desiring a thinner and a revolving stage, he begged Mr. Zentmayer to devise one and fit the same to his stand. Quite a correspondence ensued, and the army stand was thus equipped, the new stage being practically the same as the small one furnished with the Centennial. It worked nicely, and was, in truth, all that the author desired.

In the change of stages proposed, attention is called to the fact that the smaller stage has not the graduated edge. This to the majority of users would not be a serious objection, while, on the contrary, many would gladly avoid the cost of the graduated circle.

Mr. Zentmayer makes some eight or nine different forms of stands. We have room to describe only one other, viz.:

THE AMERICAN HISTOLOGICAL STAND.

The requirements held in view, in the construction of this little instrument, were the combining of the facilities of a first-class stand with moderate cost.

The entire instrument is made of brass, the base and uprights are one piece, of a peculiar shape, and of great rigidity, to which the bell-metal bar is attached by a

joint, allowing the use of the instrument at any angle of inclination; perpendicular and horizontal positions are indicated by stops; the coarse adjustment is accomplished by a sliding tube, the tube being but five and one-half inches long, but capable of elongation to the standard length.

The fine adjustment is similar to that of the Centennial — a concealed lever moving the entire body; this adjustment is reliable and very delicate. The sub-stage, plane and concave mirrors, swing in the same manner as do those of the Centennial, having the object in its centre, even when swung over the stage.

The sub-stage carries the diaphragms, of which three are furnished with the instrument. Any piece of sub-stage apparatus, such as condensers, paraboloids, prisms, in fact anything from the lists, can legitimately be adapted to the sub-stage, or the same can be instantly removed, with the mirror also, if desired, thus leaving the stand free from any obstruction below the stage.

The sub-stage slides up and down in strong dovetailed grooves, and has centering adjustments by hand. The weight of this little stand I judge to be about three or four pounds; it can, on a pinch, be carried in one's great-coat pocket.

It is worthy of mention in connection with these two stands of Mr. Zentmayer's, that the fine adjustment has been removed from the front to the rear, or, as Mr. Zentmayer says, "to the more stable part of the instrument." The author, when his attention was first called to these stands, regarded this change of the fine adjustment as one of doubtful utility; and he once made the remark to a friend, that should he purchase a Centennial, he should insist that the maker furnish a fine adjustment, both front and rear.

Certain it is that, in the process of correcting a first-class wide-angled objective, it is a convenience to have the fine adjustment at the front. It is, however, exceedingly difficult to so accurately fit the sliding nose-piece that it shall move up and down with perfect free-dom when acted upon by the fine screw, and at the same time to be free from *lateral* motion, and this lateral motion constitutes a first-class fault to which the attention of the reader has been called on another page. The very best makers have found it difficult to steer clear of this. The writer once met with one of Mr. Zentmayer's larger stands exhibiting this defect in a marked degree. His own " Grand American Stand," however, in this respect, is faultless.

Thus it will be seen that, in the nature of things, a really fine adjustment acting on the nose-piece involves skilled labor, and this, in turn, involves cost.

Again, the movable nose-piece necessarily changes the length of the body-tube, and this, in turn, again, changes the amplifying power of the objectives, and to get rid of this is indeed " a consummation devoutly to be wished."

Having had the little Histological in constant use— *i. e.*, from six to ten hours daily—I am now prepared, from my own experience, to state that I am quite well satisfied with the fine adjustment as placed by the maker. It would probably take me a trifle longer to adjust nicely an objective on the Histological, and but a trifle; while, on the other hand, all the objections connected with the adjustment at the nose-piece are avoided.

The Histological, as furnished by the maker, has simply a plain stage with spring-clips. This defect did not pass unnoticed by the writer, Mr. Zentmayer responding promptly to his request with the improvised stage described on a preceding page.

It remains to be noticed that the Histological has neither rack nor pinion,* and that the coarse adjustment is effected by sliding the body within an adjustable "jacket." There is no novelty in this, for the sliding tube is "as old as the hills," and has been extensively adopted in the construction of cheap stands.

The author felt very much like kicking at this feature of the Histological. In a little time, however, experience taught him that the sacrifice of the rack and pinion was not such a serious matter as might be supposed. The sliding movement of the body tube within its jacket in the little stand is very smooth, regular, and reliable, while, on the other hand, there are some advantages accruing to the slide that are not to be obtained by the use of the rack and pinion.

For example: suppose we are working over wet preparations, and unfortunately the front of the objective becomes immersed in the liquid—a misfortune

---

*Mr. Zentmayer now furnishes the Histological, with or without the rack and pinion coarse adjustment.—AUTHOR.

liable to occur daily. It is then, in such cases, a *positive convenience* to be able to pull the body-tube out of the jacket, cleanse the objective, and return to its place. All this can be done in much less time than would be required, were the instrument furnished with rack and pinion, to unscrew the lens, cleanse, and screw in place again.

The greatest objection that the writer has to urge against the Histological stand is this: When it is placed in a vertical position, there is a liability of its tipping forward. This can be prevented by clamping the instrument to the table with a small iron clamp, such as are used by carriage-builders and are sold at the hardware shops for a dime. In this simple way the stand is rendered as solid, more stable indeed than those of the heaviest build.

One of my correspondents, to whom I had stated the above-mentioned objection, informs me that Mr. Zentmayer has made some changes in the foot, securing thereby greater steadiness.

Previous to the introduction of the Histological, it was generally taken and accepted that the purchaser of a cheap stand ought not, in the nature of things, to expect an instrument capable of all work. To a slight extent, and to a slight extent only, does the remark hold good to-day, for the Histological has not the revolving platform, for the measurement of angular apertures (apertures *can*, however, be measured on the stand), nor has it the circular graduated stage; nevertheless, the Histological will accomplish a larger variety of work than can be performed on the "Grand American Stand" of the same maker, and this, too, be the character of the work what it may, be it the study of a Histological object or the display of the No. 20 of the Moller plate, or the 19th band of Nobert.

As may be arrived at by the tenor of the preceding remarks, the author regards the introduction of the histological as marking an era in the progress of microscope stands. The long-sought problem has been solved, and in the Histological we have a cheap, reliable, and universal stand, suitable for almost any work which may be required, and capable of carrying any and all of the various accessories which in the past have been supposed to pertain only to the heavier and (so-called) first-class instruments.

Furthermore, the author desires in this place to put on record his unfaltering opinion that, in the devising, construction, and introduction of the Histological stand, the maker has bestowed a greater boon on the "body corporate" of microscopists than has been accomplished by others, either at home or abroad. If there be an error in this statement, "time, with its revenges, will set it forth."

TOLLES' LARGE "BB" MICROSCOPE.

This instrument was designed to meet the requirements of the scientific investigator. The instrument,

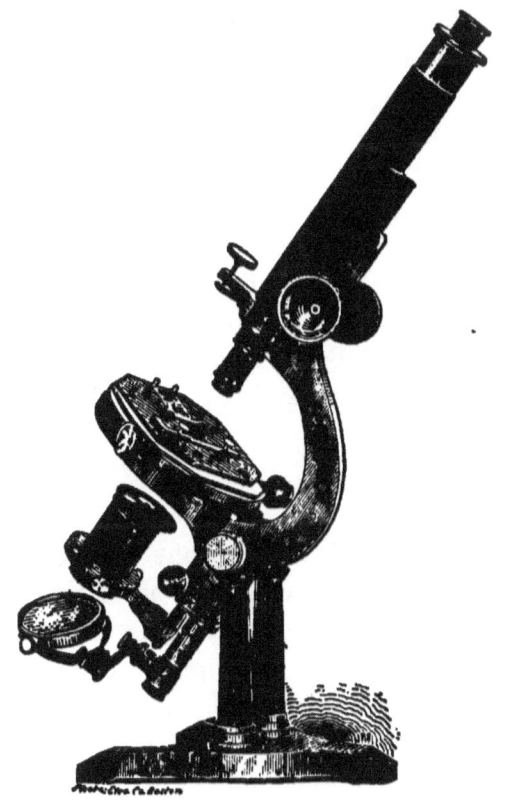

constructed on the Jackson model, is eighteen inches
high, when placed in a vertical position, and weighs
about fourteen pounds. The curved arm is supported
on a steel trunnion between two strong brass pillars

made for durability, and not liable to get out of order, and is provided with means of compensation for wear.

It has rack and pinion for coarse and micrometer screw for fine adjustment, the latter being placed in front of the body, as has been usual in first-class instruments, on the Jackson model. It is furnished with graduated draw-tube; sub-stage with rack and pinion, and centring screws for accessory apparatus; plane and concave mirrors on double-jointed arm; Tolles' thin stage, admitting light of great obliquity; with rectangular movements by screw and rack and pinion, and rotation on the optical axis of about 270°.

Mr. Tolles makes a modification of this sized stand, the stage being carried by friction rollers, and having entire rotation on the optical axis. The cost of the instrument is thereby somewhat enhanced.

TOLLES' LARGEST " A " MICROSCOPE

weighs twenty pounds, and is one of the largest and most solid instruments extant. The stage is six inches in diameter, and makes a complete revolution on the optical axis. The whole instrument rotates on a stout plate graduated to degrees, and is similar in all respects of style and construction to the " B " stand.

Either of the two stands named, made by Mr. Tolles, may be unhesitatingly pronounced first-class. The workmanship is of the very highest order; the circular stage can be so nicely adjusted as to allow of an entire revolution, under a one-twenty-fifth objective, without the object being sensibly displaced. Either form of stand can be fitted with radial arm to carry accessory apparatus at any angle. Any thing that Mr. Tolles makes is sure to be made well.

TOLLES' STUDENT'S MICROSCOPE.

This stand is fifteen inches high, weight six pounds; the base, uprights, and curved arm are of iron, hand-

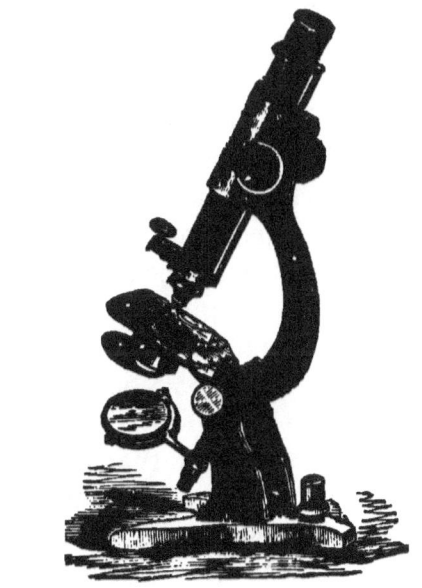

somely japanned. On a trunnion joint, made on a plan to wear well, the instrument can be placed in any position, from vertical to horizontal, and has a stop to prevent movement in either direction beyond these points. The stage is plain, with spring-clips for holding the object slides; revolving diaphragm; concave mirror, with movement to give oblique light, and for the illumination of opaque objects the mirror is removed to an upright stand. The coarse adjustment for focus is effected by sliding the compound body which is held in its place by a spring; fine adjustment by a movable plate and screw on the stage, which is efficient with high powers. The stand is made with all the care bestowed on his first-class instruments. The form is that of the Jackson pattern. To this instrument Mr. Tolles supplies several variations and additions, as a matter of course increasing the cost as well as its capacity. Among these several extras may be mentioned sliding stage, giving vertical and horizontal motions by hand, and adapted to the use of the " Maltwood Finder;" sub-stage for accessory apparatus; fine adjustment by lever and micrometer screw; rack and pinion for coarse adjustment; thin glass stage to rotate on the optical axis; the stand entirely of brass, etc. For a "student's stand" this is an instrument of good round proportions. It stands firmly on its legs, and the stage is remarkably roomy. The body-tube is nickel-plated, and the entire instrument symmetrical in its proportions, and not without pretensions to style. Like all of Mr. Tolles' work, it is made " for keeps." There are many of them in use and doubtless giving satisfaction.

The Bausch & Lomb Optical Company manufacture several excellent stands, which were designed under the immediate superintendence of Mr. Earnest Gundlach. The firm furnish some seven or eight models, from which we select the following three:

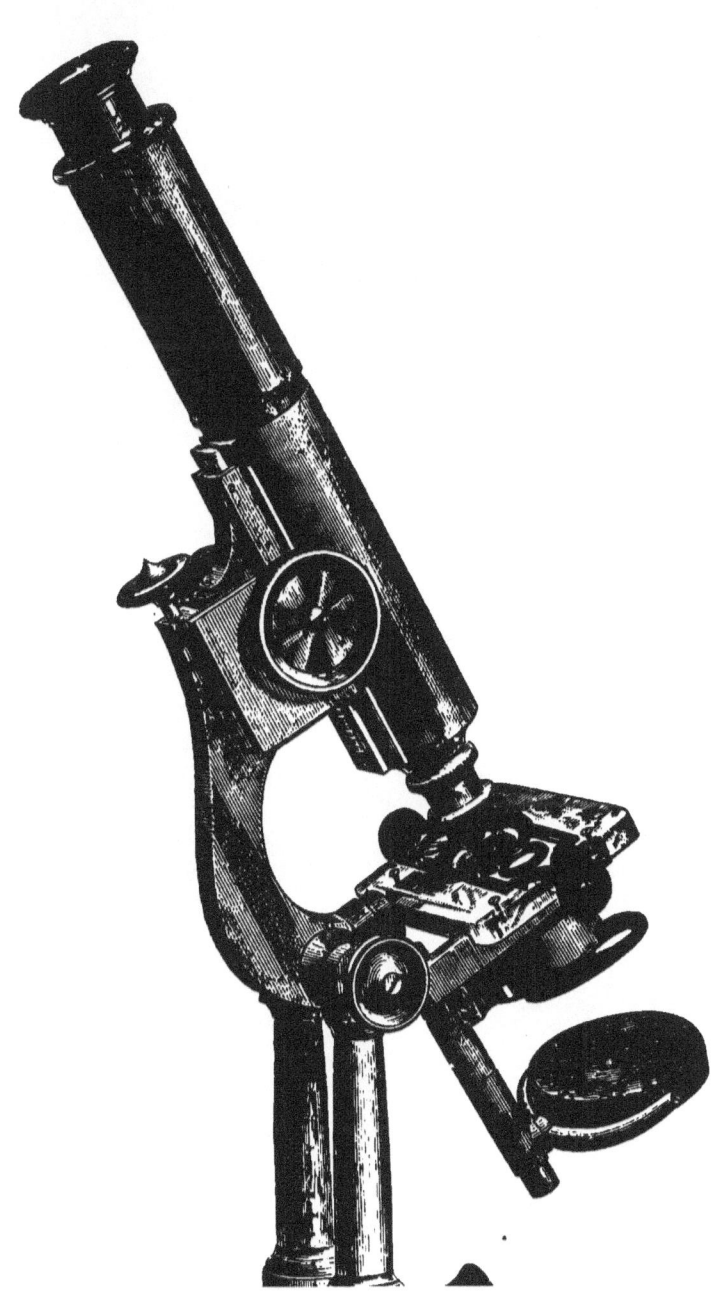

THE PROFESSIONAL MICROSCOPE.

This is the largest and most expensive instrument`
made by the company, and may be described as follows:
Heavy brass foot and pillars, both highly finished,
carrying the axis for inclination of the body, which
movement can be easily tightened or loosened by two
strong milled-head screws. Coarse adjustment by rack
and pinion, moving along prismatic slide, of first-class
workmanship, attached to the body; fine adjustment by
a new and patented frictionless motion. The object
slide rests upon a newly devised carrier; the body tube
has an inner draw tube, with society screw to which objectives of very long focal distance can be attached; large
plane and concave mirrors; sub-stage for receiving accessories of standard size, and two revolving diaphragms,
one of the latter belonging to the condenser; all attached
to the swinging mirror bar, the axis of which is placed on
the level of the object so that the diaphragm and mirror
swing concentrically around it. The mirror as well as the
sub-stage can be moved on the mirror-bar to and from
the object, and both can be removed, the latter by a horizontal prismatic slide. The sub-stage ring is provided
with internal "society screw" for objectives, condenser,
etc. There are also two slot diaphragms of different
widths, covering the whole surface of the mirror, and
only allowing light to pass through the slot in such a
direction that very sharp shadows by oblique light will
be produced.

LARGE STUDENT'S MICROSCOPE.

Heavy japanned cast-iron foot, with highly finished brass pillars, carrying the axis for inclination of the body; brass arm; coarse adjustment by rack and pinion; fine adjustment by a new and patented motion. The special advantages claimed for this new adjustment are: (1) exceedingly easy and smooth movement of the fine screw both ways; (2) perfect freedom from all lost motion; (3) perfect freedom from any side motion of the image; and (4) extraordinary durability.

The microscope is provided with a movable slide holder, serving as a substitute for a mechanical stage. This slide holder consists of a German-silver plate of very light weight, moving on a strong glass plate which forms the immovable stage; only four small points of the German-silver plate touch the top of this glass stage, while two prolongations of the former, bent downward and backward, and acting as springs, press against the underside of the glass plate with just sufficient force to keep the slide holder in position, and to prevent it from slipping off when the instrument is inclined. Two small knobs facilitate the handling of this slide holder, and it is claimed that this arrangement exceeds in smoothness and evenness of motion the movable glass stages commonly used, while the movable part has less weight, and allows the glass plate to be of sufficient strength to guard against easy breaking.

The instrument has also plane and concave mirrors; sub-stage of the extra size required to receive standard sized English accessories; revolving diaphragms, etc. These are all attached to the swinging mirror bar, the axis of which is placed at the level of the object, so that the diaphragm and mirror swing concentrically around the object. The mirror can also be moved on the mirror bar to and from the object, and the distance between the latter and the sub-stage can be varied by reversing it. Both sub-stage and mirror can also be removed.

It is to be observed that in these two instruments the importance of the swinging bar, before described in connection with the Centennial and Histological stands of Mr. Zentmayer, has been recognized by Mr. Gundlach. The mechanism, however, by which these makers accomplish the swing in the plane of the object, is by no means the same in their respective stands. Thus the peculiar method of securing the stage to the limb, employed by Mr. Zentmayer, allows that the sub-stage with its accessories, and the mirror, be brought entirely above the stage, the only thing preventing a *full* revolution being the body tube, while in the instruments of Mr. Gundlach the swinging-bar plays in a slot cut in the rear of the main stage plate, and the swinging motion is thus circumscribed by the stage slot.

FAMILY MICROSCOPE.

This has japanned cast-iron foot and pillars supporting the axis which carries the body, so that it may be inclined to any angle revolving diaphragm below the stage; rack and pinion for adjustment of focus; concave mirror, adjustable for oblique light; plain stage with spring clips.

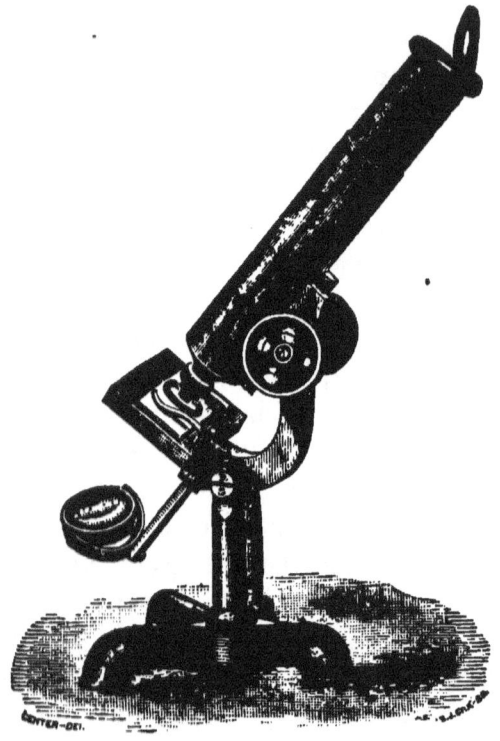

This, although the least expensive instrument made by Mr. Gundlach, is nevertheless well put together, strong and really serviceable, and, with proper care, ought to last a lifetime. It can be readily made capa-

NOTE.—Since the above was written, we learn that Mr. Gundlach has separated from the B. & L. optical company, and that he now devotes his entire attention to the manufacture of object glasses, the Messrs. Bausch & Lomb, however, continued to furnish both stands and objectives, and are by purchase, the proprietors of many of Mr. Gundlach's patents.

ble of doing most of the work required by the physician and to those who can afford the luxury of having two stands something of this kind will be found a real convenience.

Mr. W. H. Bulloch, of Chicago, manufactures seven different forms of stands, from which we present the following:

BULLOCH'S FIRST-CLASS MICROSCOPE A 1.

It has the concentric, rotating, and mechanical stage, with graduations for measuring angles; is also adjustable, so that it can be accurately and perfectly centered. There are also graduations connected with the horizontal and vertical movements of the stage, by which the exact position of an object can be noted and found with more certainty than with the "Maltwood Finder." The whole stage is sufficiently thin to admit an angle of oblique light as high as 134 degrees, and, if required, the stage can be made reversible.

The sub-stage is fitted with the most complete movements for centring or for oblique light with the achromatic condenser. It has one-fourth inch movement each way; rack and pinion; divided circle for polarizing; is so arranged that the sub-stage can be used either above or below the main stage, and can be operated by hand or by tangent screws; it is entirely separate from the mirror, but if desired can quickly be connected, so that the mirror and sub-stage turn together around the same centre, which is the thickness of an average slide over the stage.

The entire sub-stage with its milled heads can be taken off, so that there shall be nothing in the way when using direct light. The mirror is arranged so that it can be used in any direction, backward, forward, over or under the stage. The mirror and also the sub-stage have graduated circles, so that the obliquity can be noted.

The movement of the body is effected by rack and pinion connected with two milled heads, which connect with the lever of the slow motion, thus preserving the distance between the objective and the eye-piece. The slide on the body tube is long and broad, thus preventing vibration or lateral motion when using the milled heads or the micrometer screw; the latter is grooved so that it can be used for photographic purposes.

The instrument is mounted on tripod base, with revolving platform. The platform is graduated, and upon it two standards are fixed, between which the instrument turns to the angle when so used, or turning horizontally for drawing, or for measuring angular aperture. The centre on which the instrument turns, when placed horizontally, is in a direct line with the object on the stage.

The binocular model is arranged with rack and pinion for different width of eyes; the prism is so fixed that the distance remains the same. The society screw at the end of the body is arranged as a safety nose-piece, with spring, so that the danger of breaking slides is

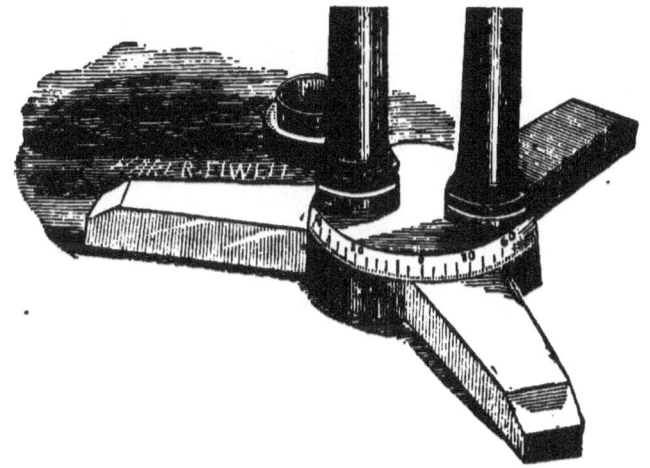

avoided. The iris diaphragm has the society screw so
that any objective can be used for a condenser, or it
can be used above the objective as an adapter, to reduce
the light in the instrument. The stand is nineteen
inches high when arranged for use.

This stand and the " Centennial " of Mr. Zentmayer,
may be considered as rival instruments, and neither of
them enters the list for a slow race; both are respectively the masterpieces of their makers, and are claimed
to be the very embodiment of all and singular that can
be desired by the microscopist.

Purely in the interest of the reader, we proceed to
compare these stands with each other; the final verdict
shall be left a matter of individual election.

First. The price of both instruments is precisely the
same, while the Bulloch has additional the iris diaphragm, the mechanical stage with its adaptations for

use as an object finder, the safety nose-piece, the duplex
arrangement of sub-stage and mirror arm, whereby
these move in unison, or independently of each other,
and the centring and rotating sub-stage.

To meet this, it might be claimed, on the part of the
Centennial, that the above-named additional contrivances have been purposely discarded by the maker of
the Centennial; that really there is no particular advantages to be derived from the iris diaphragm, the mechanical stage, the centring sub-stage, or the duplex
arrangement above referred to, and that all these contrivances serve unduly to complicate the instrument.

As to the centring sub-stage, it might be also said
that with the introduction of the "swing" there is no
longer need of wide-angled achromatic condensers.
Furthermore the secondary body of the Centennial
being itself accurately centred, there is no occasion to
introduce especial appliances for this purpose.

It will be noticed that in the Centennial the mirror
and sub-stage rise without hindrance above the stage,
the movement being only stopped by contact with the
body tube (supposing the smaller stage to be employed,)
all this time the mirror remaining in fixed focal position. To accomplish the rise of the mirror above the
stage in Mr. Bulloch's stand, the jointed arms connected
with the mirror are brought into play, and indeed cannot be dispensed with. A close comparison of the
mechanism will reveal the fact that in respect to the
swinging arrangement, there is a wide difference in
the construction of the two instruments. The angle of

obliquity obtained with the Centennial stage, is greater than that obtainable in Mr. Bulloch's stand.

The principal points are thus presented; it remains for the reader to use his own election. Either of the two stands will bear comparison with any foreign stand extant.

Mr. Bulloch also supplies another stand, which he calls his

SMALL BEST STAND "A–B."

This instrument is similar in construction to the large stand, but smaller, excepting the body tube, which is of the standard length in all his instruments.

MR. BULLOCH'S "D" STAND.

This has been selected by the author for description principally because it is a medium-sized "student's stand," and furnished with a concentric adjustable stage, it being desirable to present as much variety as possible within our circumscribed limits.

The adjustable concentric stage can be centred to any objective; the edge of stage is bevelled and graduated, so that angles of crystals can be measured; it has a complete revolution. Its glass stage is perforated in the centre, has brass fittings, so that the "Maltwood Finder" can be used; and its motion is perfectly smooth under the highest power. To the underside of stage is fitted a tube for accessories; this can be removed so that the utmost angle of oblique light can be obtained. Plane and concave mirrors and lateral motion for ob-

lique light, coarse adjustment by rack and pinion, fine
ditto, by delicate screw, diaphragm; draw tube graduated, with society screw at end. The instrument can
be inclined to any angle, has English horseshoe base,
and is sixteen inches high when arranged for use.

BULLOCH'S NEW BIOLOGICAL STAND.

The cut shows the instrument about two-fifths the
real size, it stands twelve and a half inches high, and
the stage three and a half from the table when in an
upright position. Body tube five inches long, draw
tube five inches long; with marked ring when drawn
to the standard length. Standard size and sub-stage
fitting, adapted in sub-stage with the society screw for
the use of objectives as condenser, and the diaphragm
fits in the same screw and can be used close up to the
object slide. Plain and concave mirrors each swing
over the stage can be used together, or separate, can
be clamped in any position, by milled screws shown
behind the limb. Spring 'stop to mirror and sub-stage
when in line axis. Concentric revolving stage, can
be adjusted concentric to the axis by means of capstan-
head screws, the sub-stage also can be adjusted concentric by means of capstan screws. Spring clips to
stage or the sliding glass stage as desired.

When the stage is not required to revolve, it can be
clamped in any position by the milled screw shown in
front; and when there is any danger of injuring the
stage by the use of acids, the stage lifts out of the ring
in which it revolves, and any ordinary piece of glass

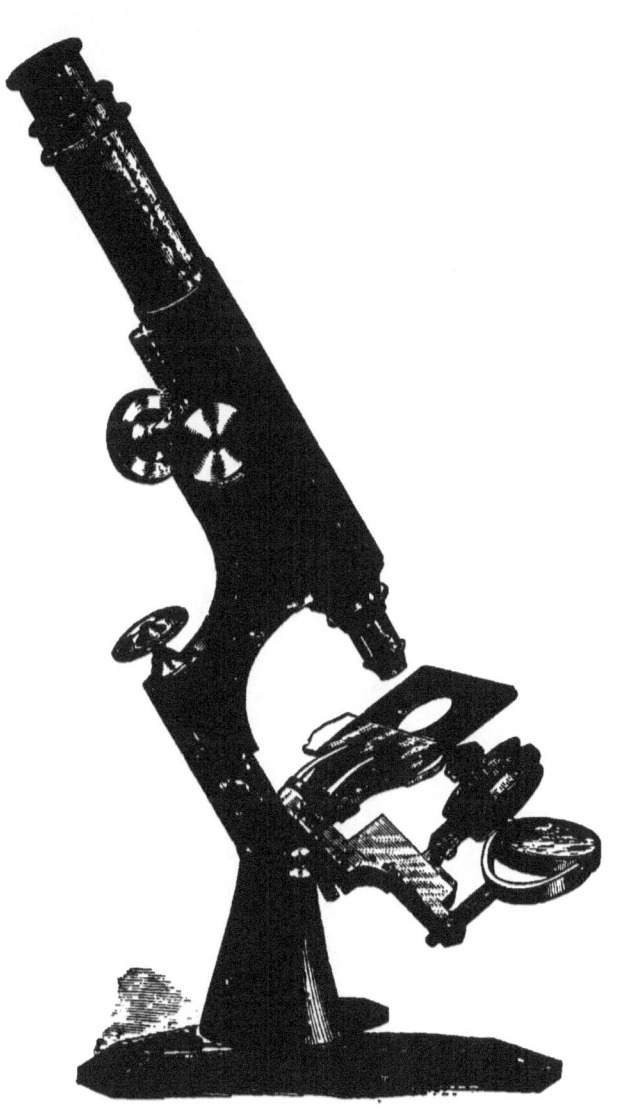

can be used on the ring by placing the instrument in an upright position. Rack and pinion quick motion, lever fine motion placed behind the limb moving the whole body tube; the body tube is fitted with broad guage screw one and a quarter inch diameter, and in which fits adapted with regular society screw, tripod base with single pillar, the axis on which the instrument turns is placed in such position that the instrument is perfectly balanced when placed in horizontal position for drawing. When so placed the centre of eye-piece is seven and a half inches from table, the standard is furnished with a B eye-piece. The mirror and sub-stage can be fitted with divided arch for measuring the obliquity of light, the stage can also have divided circle for measuring the angle of crystals. Stand all made of polished brass.

R. & J. BECK'S MICROSCOPE.

R. & J. Beck, of London, the old and well-known firm, have established an American agency in the United States. In their facilities for the production of first-class workmanship they are not excelled. As makers of London instruments, they have been long and favorably known, and the Londoner points with commendable pride to the names of Ross, Powel, Leuland, and R. & J. Beck.

No attempt can be made in these pages to describe a tenth part of the stands manufactured by the Messrs. Beck. Their catalogue will be found to include all the various grades, where may be found elegant and expensive stands, which, with the accompanying accessories, etc., cost $1,600 and upwards, down to the more simple forms, within the reach of all. It is from the latter that we make the following selections:

BECK'S POPULAR MICROSCOPE, MONOCULAR OR BINOCULAR.

In this stand the arrangement for changing the angle of inclination of the body is new and durable. It has coarse adjustment by rack and pinion, fine adjustment by micrometer screw; the stage is fitted with improved object-holder and concentric revolving fittings; the mirror slides on the main stem, and has its own semicircle for universal motions. A diaphragm is provided with perforated revolving disk. The instrument is on the " transverse arm " model, which has been so extensively used by Ross, of London, as well as others.

The instrument can be furnished with all the usual accessory apparatus as supplied by the makers; a mechanical stage, giving horizontal and vertical motions by screws, is also furnished if desired.

BECK'S ECONOMIC MICROSCOPE.

The makers say in their catalogue that " the microscope is now such an absolute necessity for the student, to enable him satisfactorily to carry on his investigation, that it is more than ever incumbent on the optician to construct a sound economic instrument adapted to the special requirements of this large and increasing class."

To this proposition let this little book say, Amen.

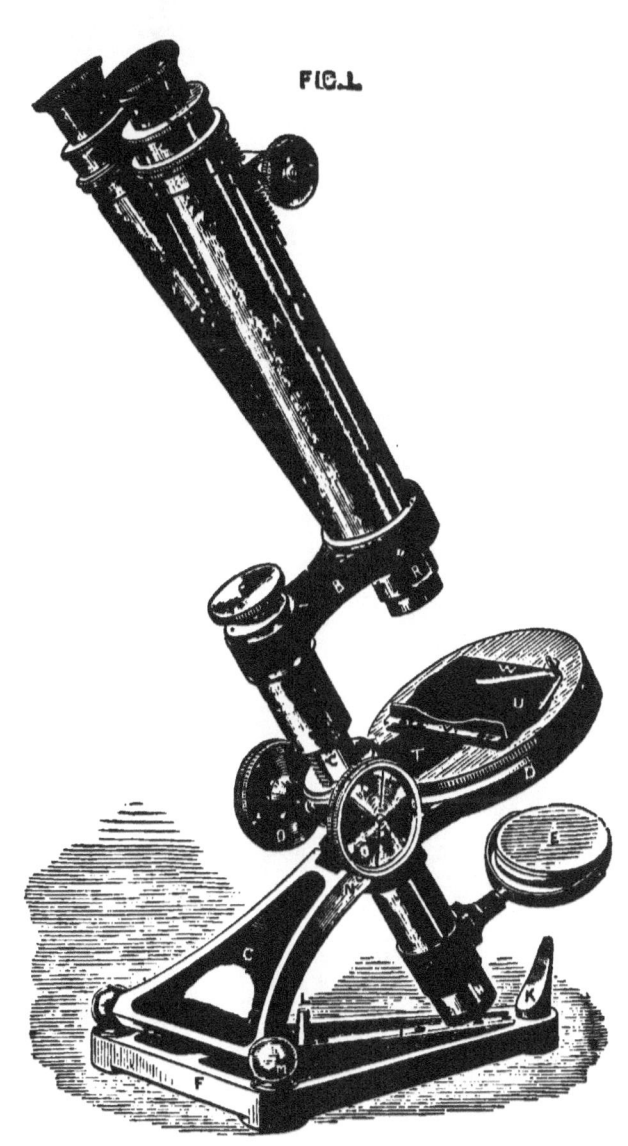

FIG. 1

"For ordinary pathological, physiological, and botanical investigations, many of the delicate adjuncts applied to the higher-priced instruments are unnecessary, and tend rather to confuse than to assist the beginner."

To this let these pages say again, Amen.

"A firm stand and well-corrected object-glasses are, however, indispensable."

In response to this we not only say amen, but shall, in the proper place, have something further to add.

"The stand of the "Economic" is made in two forms: the one with a sliding adjustment for focussing the object, and the other where the quick movement is produced by rack and pinion. On both stands the fine adjustment is given by means of a milled head at the top of the stem. The stand is fitted with the society screw.

The foundation of the stand is a heavy horseshoe base A, at the head of which is a firm pillar B, having at its top a hinge joint C, which allows the body D to be inclined to any angle, and is sufficiently firm to permit of its being placed horizontal for use with the camera lucida.

The body tube is short, but by means of the lengthening draw tube V can be made of the standard length.

In the more expensive stand the coarse adjustment is by rack and pinion; in the cheaper instruments the quick movement is produced by sliding the body D up and down the tube H, sliding over the inner stem with a spring inside, the fine adjustment being accomplished by the milled head I.

The stage K, upon which the object is placed, has two springs LL, the pins attached to which may be inserted in any of the four holes on the stage, and by their pressure, which can be varied, they will hold the object under them, or allow it to be moved about with the greatest accuracy.

The mirror M, besides swinging in the rotating semicircle N, attached to a bar O, with a joint at each end, allowing a lateral movement, so as to throw oblique light on the object, and for this purpose the tube beneath the stage, carrying the diaphragms, has semicircular openings cut on either side, leaving a clear and thin stage, allowing the utmost obliquity of illumination. This tube also carries the polariscope, etc.

The diaphragm P slides in the sub-stage fitting, and consists of a tube containing two caps furnishing two sizes of openings, immediately in contact with the under surface of the slide to be examined, and also completely cutting off all light from the mirror, when opaque objects are to be viewed.

The instrument packs, without being taken to pieces, in a small and neat case, is very convenient for traveling purposes, and entirely adequate for very many purposes. The general workmanship seems in keeping with the reputation of its makers, while its cost places it within the reach of all. At an advanced cost, it can be supplied with all the usual accessories.

Last, but not least, I now have the pleasure of presenting to my readers

BECK'S NEW HISTOLOGICAL DISSECTING MICROSCOPE.

This instrument combines a compound microscope with a single and dissecting one in a very compact and

practical form. The stout immovable arm carrying the lens when used as a single microscope, is so arranged that a compound body with *eye-piece* and *draw tube* may be attached to its upper surface, whilst beneath it is

fitted with the *society screw*, whereby an objective
may be used with it. The rack-and-pinion adjustment
work so smoothly that a one-fourth inch objective may
be focussed with exactness. The mirror beneath the
stage is so adjusted upon a swinging arm that it may
be turned up *over* the stage for the illumination of an
opaque object. A revolving diaphragm, with various
sized openings, is attached to the under side of the
stage. The outfit consists of a single lens of one-inch
focus for dissecting and botanical work, and an achromatic objective of one-fourth inch focus, the same as
furnished with the economic microscopes, and one eye-
piece, giving a range of powers, with the draw tube,
of between 200 and 300 diameters, a pair of brass pliers,
two dissecting needles in ebony handles and a glass
plate with ledge. The whole packed in a neat mahogany case with lock. Several accessories are applicable
to this instrument.

THE NEW NATIONAL MICROSCOPE.

The stand which is fifteen inches in height, is constructed entirely of brass, of the highest finish and best
workmanship, having a broad, heavy *tripod* base. From
the centre of this base rises a stout column, to the top
of which is attached, by a firm joint, the *Jackson model*
arm, carrying the compound body, by which the inclination can be varied to any degree, from vertical to horizontal, the whole instrument being perfectly steady and
free from tremor in any position. The very highest
powers may be used with it, as the body, being sup-

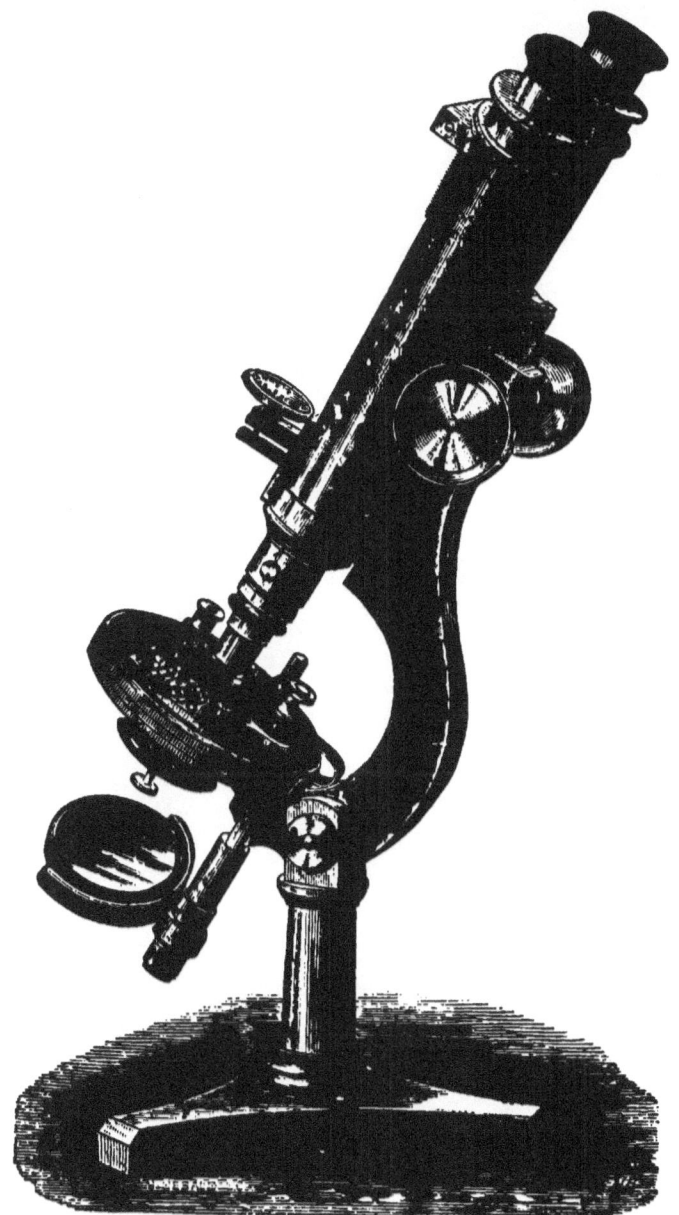

ported by the arm throughout its entire length, cannot have any unsteadiness or motion of its own.

The quick adjustment of focus is effected by means of rack and pinion, with large milled heads, which works so smoothly that there is no need to use the fine adjustment for any power lower than one-quarter of an inch. The latter adjustment is by means of a delicate micrometer screw and lever attachment, working with absolute freedom from all motion, and by which the very highest powers may be focused with the greatest exactness.

The stage is of glass, with a complete rotation in the optic axis, upon the top of which is a sliding object-holder, very thin, and with a spring clip for holding the object in place during rotation. This clip is removable, in an instant, and the stage forceps can be inserted in its place, thus allowing the latter to be moved about with the object-carrier. Beneath the stage is a tube carrying all the sub-stage apparatus, as the achromatic condenser, Wenham's parabola, polarizing apparatus, etc., etc. This is securely attached to the stage by a bayonet-catch, and can be instantly detached, leaving a very thin and unobstructed stage for oblique illumination. The *shutter diaphragm* is of novel construction, with the various-sized openings almost in contact with the underside of the object under examination, a great improvement upon the old revolving disk diaphragm. A double mirror *concave* and *plane* is hung upon a swinging bar, and with every possible motion for direct and oblique illumination.

THE NEW ACME STAND.

During the session of the "congress" of microscopists at Indianapolis in 1878, Mr. John W. Sidle, of Philadelphia, and myself formed ourselves into a committee of *two* for the purpose of devising a new microscope stand.

Mr. Sidle and myself were agreed in our opinions that notwithstanding the recent improvements which of late years had obtained, that there was still room for further effort in the construction of microscopes.

Is was then and there proposed, deliberated on, and agreed, that we would unite our energies in the endeavor to construct a microscope stand, which should combine every possible good, be equal to any and all work, exhibiting all the latest appliances, and withal, to combine *really reliable workmanship*, at the lowest possible cost, in fact at figures no higher than those commanded by inferior instruments.

This arrangement contemplated that the author should furnish all the suggestions which his long experience with the microscope might afford, while Mr. Sidle was charged with the mechanical part in the construction of the new stand.

This agreement has been carried out to the letter, and from the date named, until August last, Mr. Sidle and myself have been in close correspondence, and for the purpose named.

The first stand built under our compact was received by me from Mr. Sidle early in August last, and as a matter of course, it got a pretty severe overhauling. suffice it to say that I became so much pleased with our new bantem, that I gave it the name it bears, and have no hesitancy in recommending this joint production of Mr. Sidle and myself to my friends.

With this preliminary statement, I proceed to give the reader a brief description of our new stand.

The Acme with closed tube, when in the vertical position, is about fourteen inches in heighth, and its weight about five pounds.

The foot is of cast-iron; "horseshoe" shaped and similar to the Hartnack patterns, combining in the smallest compass the necessary weight and accompanying solidity. For real work in the laboratory this form of foot is believed to be superior to all others. For those who prefer the tripod model, Mr. Sidle will furnish this form at the same cost, or in finished brass at an additional cost of $3.50.

The entire instrument is supported by a heavy single pillar of solid brass, the lower portion of which passes through the cast-iron foot, the two parts being held firmly together by a clamp-screw underneath the foot. By virtue of this arrangement, the foot can be detached almost instantly, or, by a half-turn of the lower clamp, the foot can be reversed on the axis of the pillar, thus ensuring the greatest stability when the stand is in the horizontal position. The lower shoulder of the brass pillar passing through the foot is accurately turned and fitted, and when desired the moulding at the part of the brass pillar immediately adjacent to the top of the foot can be divided to degrees for the measurement of angular aperture, my own stand is thus arranged.

Besides those already mentioned other advantages occur, from having the foot of the stand removable, some of which may be enumerated as follows:

First, by detaching the base, the instrument can easily be carried in the pocket; it is also much easier to *pack*, and occupies much less space in the packing. It can thus be carried from place to place even in the narrow accommodations afforded by a small valise, the bother of lugging about the usual microscope case being no longer necessary.

Second, Mr. Sidle provides at a small extra cost a neat black walnut base *board*, furnished with small lamp fitted with universal movements. This board has also the necessary brass fittings to receive the pillar of the microscope, which can be clamped to the board. By this arrangement all that is necessary to convert the instrument into a first-class *hand* microscope is to bring the body to the horizontal position and to adjust the object and the illumination.

With the stand as thus arranged I have repeatedly exhibited objects illustrative of my lectures to the college class of 150 students, the instrument being passed from hand to hand throughout the entire class, and returned to me everything remaining in perfect order.

To teachers this feature of the stand will be of value. It will also be found a great convenience at times when the microscope is called on to furnish entertainment in the family circle, and objects can thus be exhibited with great rapidity.

By means of a *strong* trunnion joint the body of the instrument can be inclined at any desired position between the vertical and horizontal, the requisite stability of motion being secured by heavy "cheek-blocks;" the joint has also compensation for wear.

All the working part of the Acme above the base are of solid brass, bright finished, and nicely lacquered.

The main body-tube· is one and five-sixteenth inches in diameter. This tube articulates with the limb by means of heavy "T" guides or angle pieces, thus securing broad bearing surfaces and also perfect freedom from lateral displacement. Mr. Sidle has put himself to much trouble to perfect this portion of the stand.

The main-tube is five inches in length. This is supplemented with a draw-tube, which can be drawn out to the standard length of ten inches when desired.

The coarse adjustment is by rack and pinion. The rack is well cut and durable, and the movement of the tube by means of operating the large milled heads is exceedingly smooth and entirely without "lost motion."

The fine adjustment is by a large milled head place at the rear of the limb, *and operating the main tube.* This milled head is one and one-fourth inches in diameter, and is divided into twenty divisions. It acts directly, *i. e.*, without lever, on a micrometer screw cut fifty threads to the inch, each division of the milled head representing one one-thousandth of an inch. The fine adjustment can therefore be easily made to answer the purpose of a micrometer for measuring small intervals, or for the measurement of cover glasses, etc. In this adjustment the same security against lateral displacement is provided for by means of the "T" angle-pieces, as applied to the rack and pinion movement.

The lower end of main tube carries a nose-piece fitted with the "society screw." This piece can, however, be detached so that the broad-guage objectives now in process of construction can be used on this instrument.

The main stage consists of a circular metal plate, three and one-quarter inches in diameter, firmly bolted to the heel of the limb, and in such a manner as to be *isolated from the movements of the sub-stage apparatus;* four holes are drilled through the main stage plate, and so arranged that the spring clips may be adjusted to hold the object-slide in either a vertical or horizontal position. The spring clips may also be transferred to the under surface of the stage, holding the object-slide in contact therewith, when very oblique illumination is to be employed. The well-hole is of the usual size and is provided with a standard screw-thread, by means of which the Woodward prism and other accessory pieces can be readily placed in position. The stand is thus fitted for such emergencies as required a fixed and central sub-stage, separate from the movements of the mirror; the polariscope is also nicely provided for.

(NOTE.—It has lately come to the surface that the diameter of the "society screw" is not sufficiently large to meet the requirements of the optician when low powers of the widest apertures are demanded. Mr. Sidle is now making for the author a one-inch glass having a *diameter* of over one inch, and the Acme has been designed to meet such requirements).

The main stage has also conveniences for centering. Furthermore, a solid plug is furnished which screws into the well-hole, and forming when desired, a solid stage. This plug has an "X" engraved thereon for centering purposes.

The sub-stage proper, as well as the concave mirror, are attached to a swinging bar by dove-tailed blocks, The slides having compensation for wear, the sub-stage can be centered. It is one and one-half inches in diameter and is fitted to carry any of the accessory pieces usually accompanying a first-class outfit. The sub-stage and the mirror both slide with easy friction on the swinging bar.

The swing-bar traverses the face of a circle of brass placed at the rear of the stage, the centre of this circle being in the plane of an object placed on the stage; the centre of this circular plate is solid, that is to say, it presents a solid core of about one inch diameter; outside of this an angular groove is cut therein, in which swings the heel of the swing bar, the method of attachment of the swing bar to the circular plate giving great solidity as well as firmness in motion. Thus it will be seen that the "swinger" swings *not* on a centre but *around* a centre. This part of the mechanism needs only to be seen to be appreciated.

The mirror is of course removable, and a toy candle-holder is provided to take its place for the measurement of apertures.

A supplemental circular and revolving stage plate is also furnished, which "slips" on and off instantly as

may be required. And by virtue of a nice little contrivance of Mr. Sidle's, the rotatory movement is very
smooth and nearly as central as would be expected from
stands of much higher cost.

When desired, Messrs. Sidle and Poulk furnish a
mechanical stage which slips on and off in place of the
rotary plate above mentioned; this mechanical stage
has vertical or horizontal motions to the extent of
three-fourths of an inch. This, however, involves an
extra cost of $14.

To exhaust the list of American makers would be
quite beyond the province of this book. There remain
the names of Grunow, McAllister, Pike, Queen, Schrauer,
Wales, George Wale, and others. Descriptions of the .
stands supplied by these various makers have been
omitted, partially because the author unfortunately has
little acquaintance with their work, while that of others which at times he has seen, seem practically to be
much the same as those already described, and the
reader would gain nothing by the repetition.

By way of concluding this chapter, and as supplemental to what has already been said as to the choice of
a stand, the writer would especially insist on the importance of a stage thin enough to admit a beam of
light at 70°, from axis, not that the observer will
always work with oblique light, but, *when* occasion calls
for it, the stand should be capable of responding, and,
as has been before hinted, in these latter days these
kind of calls occur more frequently than was the case in
the days of yore. Let the novice especially, then, who

proposes to purchase a stand that he will not feel compelled to sell again in a few years, in order to procure a better one; keep this point well in mind.

The author, too, unhesitatingly gives his support to those stands having the swing stage and mirror; the advantages accruing from this late improvement are very valuable, of which we make particular mention of one, to wit:

In former times, when sub-stages were practically a fixture, it became necessary, in order to secure oblique light, to employ acchromatic condensers of wide apertures and short focal distance; the instrument was costly, and necessitated the employment of accurate and expensive sub-stage fittings, and, in general, could only be employed at a great sacrifice of pains, time and attention. With the introduction of the swinging sub-stage and mirror, all this is a thing of the past; the wide-angled condenser is no longer necessary for the display of difficult structures; on the other hand, it now seems desirable to employ condensers of the lowest angles, the required obliquity being obtained by swinging the substage, mirror and condensers to the proper angle. The expensive centring apparatus not being required with the low-angled condenser, their cost and bother are thus avoided; at the same time better results are obtained, and in a simpler and more convenient way.

There are other advantages, too, pertaining to the use of the swinging stage and mirror, of which mention will be made in the succeeding pages.

Finally, let me impress on the mind of the reader

that the large and expensive stands are not at all necessary; the smaller and cheaper stands, if due care be taken in the selection, will do any—practically all work required in the progress of scientific investigation. Furthermore, keep constantly in mind the fact that a great deal of work, legitimately in the line of investigation, can only be *conveniently accomplished with the small stands.*

Where expense is no object, or even when one can afford two stands without feeling thus crippled in other directions, it will be found a luxury to have the two. In the very nature of things, it is impossible to obtain the firmness and solidity of a large and heavy instrument through the medium of the smaller stands, and occasionally this stability is a real convenience, resulting in a saving of time and labor, as, for instance, in drawing with the camera lucida. The graduated revolving platform for the measurement of angles of aperture, and similarly the circular graduated stage, with its centring adjustments, are handy things to have in the house, but cost money, and, as a rule, can be dispensed with.

And last—in this matter, far from being least—should the *res angusta domi* pinch so bitterly as to compel you for the time being to omit some necessary investment, then I say, by all means *make sure* of the necessary object glasses, quantity *and quality* included, and let the stand "sweat" until the "good time coming" shall enable you to provide a new and superior instrument. Remember, too, that the finest and most expensive stand extant, fitted to a poor objective, becomes not only a dangerous tool, but also a positive nuisance.

CHAPTER II.

During the lengthy controversy which occured between Mr. F. A. Wenham, optician to the firm of Ross and Co., of London, and Mr. R. B. Tolles, of Boston, Mass., the originator of the celebrated duplex object glasses, the former presented ideas as to the functions and nature of angular aperture which were not in strict accordance with the popular views. Since then others have done likewise, and it may be well enough to leave the question open, without attempting to answer the above interrogatory at all.

For the information of those, however, who are just entering the study of microscopy, we will undertake to tell what angular aperture *was*, or to be more definite, what it was in 1856; and for this purpose, the author selects the definition found in the Micrographic Dictionary by Messrs. Griffith and Henfrey, a work generally acknowledged as authority in matters microscopical.

The angular aperture of an object-glass is the angle measured by the arc of a circle, the centre of which is formed by the focal point of the object-glass, the radii being formed by the most extreme lateral rays which the object-glass will admit.

Thus let L, in the left-hand figure 1 below, represent the lower portion of a microscope, objective,—

a perpendicular section of the lowest combination of an object glass of small aperture.

"*a* is the angle of aperture, and *f e* the most oblique rays which the object-glass will admit. The angle is measured by the dotted arc *b* in the object-glass of

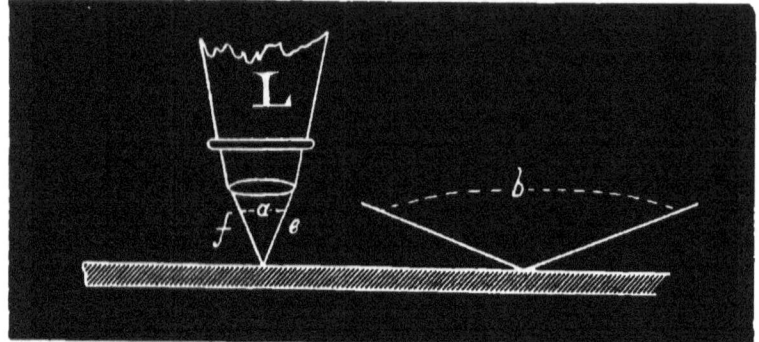

larger aperture (Fig. to the right); the arc *b*, which measures the angle, is much larger, and the radii representing the extreme lateral rays are much more oblique. Hence it is evident that the object-glass of larger aperture admits all those rays admitted by that of less aperture, and a certain number of other rays, these being more oblique."

The above definition has been and is now generally accepted. Messrs. Griffiths & Henfrey proceed to give their demonstrations of the effect of aperture, as thus defined, in an interesting and lengthy article, one that is well worth perusal by those desiring to know how things stood twenty-two years ago. Our limited space forbids its reproduction here.

HOW SHALL WE MEASURE ANGULAR APERTURE.

While the nature and functions of angular aperture remain a matter of controversy, this too may be allowed to remain an open question. Recognizing, however, the fact that object-glasses continue to be made and sold, the makers claiming for them specific apertures, we proceed to give the plan usually adopted by the optician, and at the same time adapted to the use of those working with the smaller class of stands.

It will be advantageous to the beginner to operate either in a dark room or in the evening. Select a good sized table, and on this, at the end nearest the operator, pin down a large sheet of paper, say twenty inches square. On this paper place the microscope, bring the tube to a horizontal position, and screw on the object-ive to be tested. Mark with a pencil certain points along the base of the stand, so as to enable you to describe a circle with pencil and dividers, in which the base will just revolve without lateral play. Next, provide a common candle, and, if necessary, cut this down so that the height of the flame shall be level with the objective in its horizontal position; light the candle, point the objective directly at the light, having previously removed all sub-stage incumbrances, crouching down, and with the eye at the eye-piece, adjust the direction of the instrument as nearly as possible to the candle; the field of view will of course be illuminated from the radiant.

Now revolve the entire instrument, being careful to

keep the base always within the circle drawn on the paper, and continue the movement (either to the right or left, as the case may be), until *one-half* of the field shall be bisected—*i. e.*, one-half bright and one-half in the shade. Now select some straight portion of the base, marking a straight line coincident therewith with a pencil, or, placing a rule in contact with any two salient angles of the base, draw a line.

Next, revolve the entire stand in the contrary direction, passing the candle, and until you again get a field half bisected similarly, as before mentioned. Now selecting *the same* side, or *the same* two salient corners of the stand, mark another line.

These two lines will be found divergent with each other, and with a parallel rule it will be found easy to produce parallel lines which can be continued until they meet; this done, it only remains to measure the angle obtained with a common protractor, or other instrument designed for the measurement of angles.

Where tolerable accuracy is important it will be well to cross-question the result by repeating the operation. There ought not to be a discrepancy of more than one-half of one degree.

Those who possess the large stands fitted with graduated and revolving platform will of course avail themselves of the convenience; the operation is practically the same; thus, having bisected the field (say to the right), read the angle, which will be indicated by a mark on the platform adjacent to the graduations; then bisect the field to the left, read the angle again, subtract one reading from the other, and the difference is the angle of aperture.

There being no means for the actual determination of the bisection of the fields, this having to be judged of by estimation, follows that the plan is not rigidly accurate; the author has, however, measured repeatedly the same object-glass, following as above described, and without a variation in the results of more than a half degree; many of his pupils, too, on their very first attempts, are quite as successful.

The bisection of the fields is much more easily arrived at in the evening.

OBJECT-GLASSES.

Years ago wrote Dr. Carpenter thus:

"It may be safely affirmed that the most perfect object-glass is that which combines all the preceding attributes (viz., defining power, penetrating power, or focal depth, resolving power, and flatness of field) in the highest degree in which they are compatible one with another. But, as has just been shown, two of the most important—viz., penetrating power, and resolving power—stand in such opposite relations to the angular aperture, that the highest degree of which each is in itself capable can only be obtained by some sacrifice of the other; and therefore of two objectives which are respectively characterized by the predominance of those opposite qualities, one or the other will be preferred by the microscopist, according to the particular class of researches he may be carrying on; just as a man who is

about to purchase a horse will be guided in his choice
by the kind of work for which he destines the animal.
Hence it shows, in the author's estimation, just as
limited an appreciation of the practical applications of
the instrument, to estimate the merits of an object-
glass by its capability of showing certain lined or dotted
tests, without any reference to its penetrating power or
defining power, as it would be to estimate the merits of
a horse *merely* by the number of seconds within which
he could run a mile, or by the number of pounds he
could draw; without any reference, in the first case,
either to the weight he could carry, or the length of
time during which he could maintain his speed, and, in
the second case, either to the rate of his draught or his
power of continuing the exertion. The greatest capacity for *speed* alone, the power of sustaining it not being
required, and burden being reduced almost to nothing,
is that which is sought in the racer; the greatest
power of steady draught, the rate of movement being of
comparatively little importance, is that which is most
valued in the cart-horse; but for the ordinary carriage-
horse or roadster, the highest merit lies in such a *combination* of speed and power with endurance as cannot
coexist with the greatest perfection in either of the first
two. The author feels it the more important that he
should express himself clearly and strongly on this subject, as there is a great tendency at present, both among
amateur microscopists and among opticians, to look at
the attainment of that resolving power which is given
by angular aperture as the one thing needful. . . .

It is neither the *only* nor yet the *chief* work of the microscope (as some appear to suppose) to resolve the markings of the siliceous valves of the diatomacea; in fact the interest which attaches to observations of this class, *per se*, is of an extremely limited range. . . . And the more carefully we look into the history of those contributions to our knowledge which have done the most to establish the value of the microscope as an instrument of scientific research, the more clear does it become that for almost every purpose *except* the resolution of the diatom tests, objectives of moderate angular aperture are to be decidedly preferred." *

This quotation — essentially true when first published — has been, and to a considerable extent now is, the orthodox faith of the microscopist. It has been endorsed by everybody, both at home and abroad; the world of observers have rested peacefully at ease on its broad platform; the man of science was serene and content with his medium apertures; but it was the crowning glory of the quack that a " good honest," low-angled, " working" triplet cost only a dollar or two. And it was a fact, that so far as faith was concerned, the scientist and the veriest quack were in the same boat together, and both found pleasure in endorsing the teachings of Dr. Carpenter.

It seems to me that this singular state of things can be easily accounted for. The hard-working investigators had little opportunity to study object-glasses, their time being completely occupied, and their eyes sorely

---

* " The Microscope and Its Revelations" fifth edition, 1875, page 204.

taxed in the daily routine of labor and study. Their glasses were the best that could be procured when purchased, and, as far as their personal experience with them could attest, their experience was in harmony with the laws laid down by Dr. Carpenter; and they had " no time to waste " in fussing over " diatom tests," etc.

On the other hand, the empiric replies, " Your high-angled glasses are all well enough for the diatom man, but *for my work* give me reliable French triplets of moderate apertures;" and in this statement the author is in entire and perfect harmony!

The author repeats, that ten years ago the doctrines taught, as contained in the previous quotation of Dr. Carpenter, were essentially true; he has no fault to find with their *original* publication. but he does regret that Dr. Carpenter has allowed them to retain their place, unchanged or unrevised, in his late editions.

Let us glance for a moment at the history of the object-glass as connected with its aperture. Years since it was known that the exhibition of surface-markings required to a greater or leeser extent the employment of lateral or oblique light; width of aperture hence became a desideratum, at least in such glasses as were to be used with this object in view.

Whether right or wrong, demands *were* made on the optician for objectives possessing increased apertures. Notably, this demand was first met successfully by our own countryman , the veteran Charles A. Spencer, who produced a glass of wider angle than had been previously accomplished and thereby astonishing the world, sorely punching the English opticians (who flattered themselves on secure ground at the head of the profession) with an exceedingly sharp stick; with this glass Mr. Spencer succeeded in displaying both sets of lines on the diatom, now known as *Navicula Spencerii*.

The London makers followed suit, giving especial attention to the extension of aperture, and, as might be reasonably expected, the problem was to get light of extreme obliquity *somehow* through the lens, nor were they (the makers) very particular to stand on the order of its going; hence it came about that very many glasses were made and sold having increased angle, by virtue of which they were able to display certain acknowledged difficult tests; but, on the other hand, the lateral pencils being poorly corrected (if corrected at all), these glasses could not compete by centrally disposed light with the well-corrected but narrow angles previously in use.

It was, therefore, to the wide-angled glasses of that day that Dr. Carpenter's remarks had force, the highest possible angle attainable being then limited by popular opinion to 155°. What is now known *and recognized* as a high-angled glass of 175° was then not only unknown, but would have been deemed a sheer impossibility, and it therefore becomes obvious that any remarks at the time referred to, whether by Dr. Carpenter or by other authors, cannot be applied consistently to the wide-angled glasses of the present date.

Those were the days, too, when the resolution of the

nineteenth band of the Nobert test-plate was said " to
be a matter of faith rather than of sight." The curious
reader is informed that the original of the quotation
will *not* be found in the *last* edition of the "Microscope
and its Revelations!"

Again, the high-angled objective, as at present constructed, is an entirely different affair from that of
former times. In place of achromatism, then thought
to be the *ultima thule* of perfection, the finest glasses
now made can hardly be said to be achromatic. And
then, again, the immersion system has been adopted
with its manifold advantages. Improvement upon improvement has almost marked the rolling months of
the calendar, while each succeeding year, to the number
of ten or more, has been richly laden with the fruits of
the optician.

In this wonderful march of progress, our own country, I am proud to say, has ever been at the front, and
with colors flying. As before stated, Spencer, with his
resolution of the *navicula*, commenced this march of
progress, and slackened not the pace until the reputation of American objectives were duly acknowledged
abroad. Then, indeed, he retired for a time, resting
gracefully on his well-won laurels; and now appears
Tolles at the van, who with almost superhuman genius
and energy grapples with the very laws of optics, and
bends them to his inflexible will!

Notwithstanding that the reputation of the country
was quite safe in the hands of Mr. Tolles; Spencer,
after a year or two of rest, again comes forward, assisted

by his two sons, both of them, as a matter of course, "to the manner born," and in the very prime of life and activity.

Thus far I have simply referred to Spencer and Tolles, because they have contributed largely to the production of high-angled objectives; but the list of American opticians is not thus complete; we have Wales, Zentmayer, Gundlach, and Grunow yet in reserve, whose productions in their several lines are not inferior to any imported.

In the foregoing brief and general history of the American wide-angled objective, the conclusion is inevitable, first, that during the past ten years *something* has been accomplished, and that in the interim our opticians have not been idle; secondly, that the modern high angled glass is to be judged of strictly *on its merits, and not by what was affirmed of a medium aperture ten long years ago.*

The converse of this is equally true, and the author, in presenting his views of to-day, declines to be held responsible for what the future may bring forth.

In this place it may be well to consider *what,* in common parlance, constitutes a high-angled glass. The writer has already in print stated his individual views which have since undergone no change. As a rule, reference being made to wide apertures, most persons are prepared for some great show of figures, such as "175°" or "179°"; others might call for nearly "180°"; and then again, others there are who would insist on passing the "impossible" 180° corner, and revel among the balsam angles to the tune of 100°, or even higher. In dealing with objectives of short focal distance, all this may be well enough; the writer, however, prefers to regard as high-angled, any, and all glasses, without reference to their focal lengths, which are endowed with the widest apertures obtainable. If this platform be accepted, then it will occur that a one-inch of 50° should be classed as a high-angled objective, and similarly, a two-inch of 25°. And, again, it would also then occur that a one-sixth of 130°, which fifteen years ago ranked as a wide, would now be classed as a glass of medium aperture. And furthermore it may possibly (yea *probably*) have place, that there are many observers to-day, loud in their denunciations of the "wide-angles," falling back on "The Microscope and its Revelations" for authority, who, in their habitual use of what are now known as medium apertures, are in truth the *real culprits*, to whom and for whom were Dr. Carpenter's original remarks intended.

As has been suggested on a previous page, there may be seriously some question, not only as to angular aperture *per se*, but as to what constitutes the measure of the same. It is one thing to get light through an objective, and quite another to bring said light into the traces, and render it of use to the observer. There are, too, scores of high-angled glasses (so-called) sold as having angle of 175°, and possibly more, that are entirely worthless when worked with pencils beyond 130°; indeed, many that I have seen would utterly fail when compared with a really good glass of 115°.

The writer has before him a glass of the latter angle, (115°,) made by C. A. Spencer & Sons. It is what they call one of their " professional series "—a dry one-fourth; the makers ought not to expect that this glass should be called on for work requiring oblique pencils greater than 100°.

Now this glass—this professional one-fourth—will give me real good shows, even when worked at all the obliquity obtainable on my Zentmayer stand; and thus have I seen with it the longitudinal markings on the balsamed surriella of the Moller test-plate, and this is a test that will defeat many dry eighths, engraved by their makers as having 160° or more angle.

From this little experiment—one that the writer has repeated scores of times in the interests of his friends—some curious conclusions *might* be arrived at, which, although possibly coherent and plausible in detail, become absurd when considered collectively; thus it might be held:

First. That both glasses having at least 160° of aper-ture, are in fact high-angled glasses.

Second. That of the two named, the Spencer is the better glass.

Third. That the Spencer objective has *really* but an angle of 115°, as marked by its maker; that its capa-bility of admitting working beams up to 140° gives it no real claim to those figures.

Fourth. That, as any good eighth of 140° will easily show the longitudinal markings of the surriella, it is proven that the one referred to has not that angle at all, and is, in fact, but a medium power glass—the general verdict being that it is an indifferent one at the best.

Fifth. That the capacity of an objective to display markings on balsamed and difficult tests, at or near the limits of its aperture, is no index of angular aperture. In the case presented, if it so be that the eighth, when worked at 140°, gave fair and distinct images of the surriella, then there is no reason to dispute the angle as claimed for the glass; let the experiment be repeated an using angle of 120°, and over dry as well as balsamed mounts.

The above may be taken as representing individual differences of opinion. Either of the conclusions presented have been urged on our attention time and time again. To the second, third, and fourth, the author gives his assent, and has never allowed an opportunity to pass without exposing by actual demonstration the fallacy contained in the fifth.

From the preceding, then, it becomes apparent that there is much difference of opinion as to *what* constitutes angular aperture or the measurement thereof. Until there can be some more precise plan arrived at, let the purchaser of any objective imperatively demand:

First. That the objective, in general, work with its full vim fully up to the limits of the aperture claimed for it; that the images be strong, vigorous, brilliant, and without distortion, and that such images shall not be surpassed in any particular by any similar glass, without reference to its angle. If it be that you have a wide aperture in hand, see that its work by central or centrally disposed light is not excelled by any objective of narrow angle extant. In calling on the glass at the limits of its aperture, demand, and see to it particularly, that there be no letting down of general performance, that the images remain strong and vigorous, that the corrections are not impaired, and even with the widest apertured objective known, that there be no sensible distortion of the image.

Again, if it so be that you desire thus to test an objective claiming considerable aperture, say 170° or 175°, and adjustable, note whether there be any special adjustment required when worked at or near the limits of its aperture, other than necessary for its correct performance by central light; if this be found the case, the indications are that the two sets of pencils are not in harmony with each other—as the Germans say, are not "married." It will be necessary to use a little discretion here, for some of the very finest glasses require a slight change of adjustment under the conditions named. The less of this especial adjustment, however, the better.

Any objective that will stand acceptably the foregoing tests may be allowed to "pass muster" as to its aperture *sans peur et sans reproche.*

In the act of writing the above, the author was interrupted by a friend with the remark, "Are you not screwing things down pretty fine? Don't you see that your method not only is a severe test as to working

angle, but is also a severe test as to the general qualities of an objective?"

In response to this interrogatory, I reply, that it's
high time things were "screwed down." As to the
latter portion of the remark of my friend, it may be
observed that the directions given indicate a part, *and
part only*, of the course to be pursued in testing the
performance of a really first-class American objective
of wide aperture.

The plan proposed is liable to another, and, in the
minds of some, a most serious objection; says one,
" Don't you see that you have advanced no positive
guage? Your idea simply is to compare one glass with
another — in short, it means ' fighting objectives ' "
Selah!

SOMETHING FURTHER ABOUT OBJECTIVES.

There is another matter of common acceptance which
has in the past made some mischief; I refer to the fact
that the focal or working distance of an objective has
been and is considered the index of its capacity for certain classes of work. Thus, the inch has been set apart
for the study of such objects as required examination,
with powers from 50 to 150 diameters, and where its
comparatively long working distance was desirable,
while the one-fiftieth, whose lowest power of 2,500
diameters and its exceedingly short working distance,
could not perform the work of the inch — the one-
fiftieth being reserved for the investigation of the most
minute organisms, and under the highest amplifications.

Now, in the instance cited the popular idea is correct. The inch cannot do the work of the fiftieth, nor can the fiftieth do the work of the inch, and each, as to the other, are to be appropriately brought into use.

But, as is well-known, between and intermediate to the scope of the two glasses named, there are several objectives having not only intermediate but variable focal lengths. Among these latter, too, are to be found the objectives known as " medium powers," and it is with reference to these that it may be affirmed that the broad rule governing the inch and the fiftieth does not hold good.

Some four or five years since, the author, in writing to a brother microscopist, hazarded the statement that the time would surely come when the optician would furnish one-sixths, capable of performing all the work then done with the one-fiftieth. The principal reason advanced at that date, in support of his opinion, was,

First, assuming the case of a perfect objective with a perfect eye-piece, he claimed that it made no difference to which end of the tube the power should be applied.

Second, the nearer perfection arrived at in the construction of the objective and eye-piece, the higher may be the power of the latter; and

Third, as we have no right to expect absolute perfection in the construction of objectives, it nevertheless seemed reasonable to infer that the optician could better handle and adjust a lens of sensible dimensions, such as are used in the manufacture of the medium powers, than could be possible with the merest speck of glass forming the fronts of the one-fiftieth.

Whether the ideas thus advanced were correct or not, the fact is patent that in less than two years from the date of the said letter, Mr. Tolles produced a one-sixth, that excelled for any and all work the performance of any one-fiftieth on record. This one-sixth is still in the possession of the author, who, ere the glass was thirty days old, pitted it against the finest fiftieth to be found in the country.· The battle waged for an entire week, but the result was decisive. It was David vs. Goliath, and David had the best of it.

Scarcely had another month elapsed before Mr. Tolles again sent the writer another glass — this time a tenth — which, in turn, eclipsed the previous inimitable work of the sixth; while at a still later day, Mr. Herbert Spencer produces a tenth, made on a somewhat different formula, the performance of which is not excelled by any glass yet made, " be it a fifth or a fiftieth."

Without reference to the " impossible 180°," it may be positively claimed that either of three glasses named have greater aperture than is possible (or has thus far been possible) to obtain with the fiftieths.

As has already been stated in the introduction, the writer was the first to call public attention to the claims of American objectives of medium power. Statements so radically at war with the generally accepted popular belief were destined, as a matter of course, to meet with opposition. Microscopists from almost every section came in person to see for themselves, many of them bringing their favorite high-power glasses for comparison, and returning to their homes satisfied with the trip, leaving the one-sixth and the tenth to encounter the next comer.

It being probable that there are others who yet remain to be convinced as to the accurary or validity of the claims of the " medium powers," it may be stated that at this late day the writer is no longer in a minority of one. Microscopists of note have studied the situation, arriving at similar results. About twelve months ago, Mr. John Mayall, Jr., a well-known and talented microscopist of London, wrote as follows:

" I am not going to enter into a mass of details of the various trials I have made with Tolles' one-fourth and one-eighth. Suffice it to say that no lenses that have been in my hands have ever been so thoroughly tested against the best lenses by English, French, and German opticians (here Mr. Mayall presents a list of *seventeen* recent immersion objectives by the most renowned makers in Europe); and without reserve of any kind, I say these lenses are the finest I have ever seen.   *   *   *   I affirm, then, that with central and oblique light on all the objects that are known here as tests, Tolles carried the palm. I find, on the most severe tests, there is in Tolles' lenses a better correction for spherical aberration, the image is more *crisp* and *clear*. By difficult tests, I mean, for instance, *surrirella gemma* with central light, or *amphipleura pellucida* with oblique light.   *   *   *   I urge that low-angle lenses will not exhibit the definition these lenses will show, and that if one takes a higher power that will show the images, he will find, by comparison, the higher power will be the more difficult to manage. The whole question turns upon results; if you are content with medium images, use medium or low-angle objectives; if you train your eye for fine images, you must use high-angled objectives."

In 1876, Dr. J. G. Hunt, of Philadelphia, a widely known and expert microscopist, after having given the new American objectives of medium power close study, writes as follows; and believing that the doctor's letter will be found of general interest, we give it entire. The glass he makes reference to was a Tolles tenth:

"I can now report to you that the one-tenth you sent me is grand. It contains more good qualities than are to be found in many first-class lenses—perfect mechanical workmanship, large field, gives sharp image on the margin of field, decision of definition leaving nothing doubtful or foggy, equal penetration with resolution; thus being superior for histological work. .

. . I could engrave it all over with marks of admiration. . . . For the best work of the botanist or histologist it has a definition which can be retained, with an amplification such as I have not seen in any one twenty-fifth or one-fiftieth that has come under my notice. . . . I see in its construction more finger skill, more time and conscious brain patience than mathematics. Hence its character; it has no precedent, but is wholly original, and unlike any other make, English or continental."

The preceding quotations are thus presented to the reader because the issue we have under consideration is therein discussed typically (I may say) from both the American and English standpoints. To the testimony of these talented gentlemen the author could, did his space allow, add a mass of similar evidence.

I repeat it is not consistent with the limits of this little book to further discuss the issue in question. The author therefore dogmatically asserts that his positions taken in public print relative to the matter we have been considering, were then correct, and have so remained up to the present date.

But mark this point: the claim thus established in favor of "medium powers" of the widest apertures *has no reference whatever* to hosts of objectives made and sold with high-sounding figures attached. Keep this fact in lively remembrance.

We are now prepared to return directly to the point from which we started. We have seen by our digression that the relations existing between the inch and the one-fiftieth are to be essentially modified, relatively, as to the nature and performance of a one-sixth as compared with that of a one-fiftieth. For instance, if it were true, as has formerly been accepted, that it is the province of the inch to assist in the study of the simpler organisms, and that of the fiftieth for the investigation of the most delicate structures, it does *not* hold good at this present writing that a one-sixth or tenth (generally classed as medium powers) are the proper objectives for an *intermediate* class of work *only*.

8 Microscopy.

It is obvious also that, if the one-sixth and the one-tenth are more than capable of doing the work formerly set apart for the employment of the one-fiftieth, the former have the better right to be regarded as "high powers."

Furthermore it occurs in the present advanced stage of optical science that it is really quite impossible to precisely define *what* constitutes a "medium power" glass, or for what particular class of work would such glasses (if defined) be characteristically adapted. The author confidently believes that still further and greater improvements in American objectives are yet to be accomplished. He believes, too, confidently that, as the instrument shall approach perfection, and still higher eye-pieces be brought into requisition, he may yet live to see the Nobert nineteenth band with a half-inch objective and a one-sixteenth eye-piece. It therefore seems to him that any discussion as to the characteristic duties of an object-glass based entirely on the focal length of the same may wisely be dismissed as being (for the present at least) impractical, if not impossible.

And now for some remarks that are not only *possible*, but can, if the reader elects, be made eminently *practical*, closely related, too, to what has just been written; and in these the author hopes to render some service at least to a portion of his readers.

A fine objective is in its very nature a costly instrument, while on the other hand, it often happens that the true lover of nature has unfortunately a light purse. In fact, this *whole* situation is to be regretted, and certainly to be ameliorated if possible. It happens, too, that the so-called " high powers," such as one-twenty-fifths, one-fiftieths, and one seventy-fifths, cost in themselves more than the majority of observers could afford to pay for an entire outfit, and thus have been accessible to only a favored few. The price of the fiftieth, as furnished by eminent makers may be quoted at from $250 to $300. I dare say that more than one of my readers, earnest workers with the microscope have yearned time and time again, as they have read of some wonderful things accomplished with a twenty-fifth or one-fiftieth, for the means to enable them to pursue similar investigations. Let all such hail with joy the announcement that these costly glasses are no longer a necessity, and that their work can be not only done, but better and with greater ease accomplished, with what are known as medium-power glasses of wide apertures; that there is no longer, too, any necessity of going abroad, or paying duties thereby; that a one-sixth, or, at the furthest, a one-tenth, costing from $60 to $85, will (if properly selected) compete in performance with any one-fiftieth extant—a fact, reader, worth knowing.

Another dogma in the popular mind has very general acceptance—to wit: that angular aperture can only be obtained at the sacrifice of working distance. The old saying is, that " it's a poor rule that won't work both ways;" hence it should obtain conversely, that *with* the sacrifice of working distance, angular aperture ought to be obtained; but this is not always the case. For instance, one-sixth are now made with apertures we will say (to keep out of controversy) *up to* 180°, and with a working distance of one-fiftieth of an inch. Now the widest angled one-fiftieth in existence, with a working distance less than half that of the one-sixth, will be found to measure less than 170°, and in the latter glass it is evident *that working distance has been sacrificed without corresponding increase of angular aperture.* The case cited is an instance notably in point, and one that cannot be dodged, and yet to a certain extent the same will apply to some of the intermediate objectives. Take again the before-mentioned sixth of the widest angle known, and its working distance of one-fiftieth of an inch. It will be found that, although it is possible to obtain the same aperture for the tenth, the working distance will suffer decrease; and here again is another instance where sacrifice of working distance is not accompanied by corresponding increase of aperture.

The subject is by no means exhausted, and is well worth a little ventilation. We can better get at the situation by supposing a case which might possibly occur in practice. Suppose then, reader, that you desire five one-inch glasses, each glass to have a working distance of five-tenths of an inch, and each to magnify with the two-inch ("A") eye-piece, fifty diameters, and that you gave these five glasses respectively to five opticians, to be made as per the conditions named. Now, it will most probably occur that when you get these five glasses in hand, the working distance and the magnifying powers of each are true to the specifications, and *further*, that no two of the five glasses will have the same angular aperture.

It's well enough to pause here and allow the approach of a deluge of threadbare argument. Says one, "All this proves nothing. It's quite possible that the glass with the narrowest aperture may be the better corrected. Let your own rule be here applied, and the *quality* of the apertures tested; nothing short of a competitive examination can be determinate."

To this the writer says amen; but the reader is again reminded that this is nothing more or less than "fighting objectives." This is the course, too, which the author has pursued in the way of making competitive examinations of objectives in his own interests, and in behalf of those of his pupils, his friends, and his correspondents. The result being, in nineteen cases out of twenty, that the glass with the wider aperture proved in every other respect the better glass—a result, too, not improbable in its nature, when it is borne in mind that those of our opticians who have given great attention to the development of aperture are no ways behind-hand in their general professional attainments.

Again, (to steer clear of cavil or controversy,) suppose that of the five glasses before named, all having the same working distance and amplification, the one with the lower aperture being made by Mr. X., and the other of wider angle being by Mr. Y., both objectives being, too, equally well corrected. That such a condition of things is *possible* no one will attempt to deny. Here *is* a condition to which the popular dogma *can* be applied to advantage—to wit: we can send to Mr. Y. for another inch similar to the one in hand, but with a lower aperture, and corresponding to that of Mr. X., and it will obtain that Mr. Y., in cutting down the aperture of his inch to that of Mr. X., will *increase his working distance;* and here (comparatively) we gain working distance without loss, or, as it has been termed, sacrifice of angular aperture.

Hence we arrive at the conclusion that the function recognized as "angular aperture" *per se* is not a fixed and definite quantity nor one that can be fenced in and subjected to any fixed rules. Nothing definite in the way of rigid law can be applied to it. In the case just mentioned, another curious conclusion might be arrived at, and justly too. For instance, the decrease of aperture from that of the wider aperture to that of the lower would not only be accompanied (accepting the popular dogma), which in the case in question would hold true by an increase in the working distance, but the *penetrating* power of the glass would thereby be enhanced, and this, too (comparatively), without loss of angle.

The facts presented are valuable, are significant, and worth careful thought and study. The author has never seen them in print, and they are, as suggested, the result of an active experience.

And this brings us to the consideration of another matter; I refer to the popular dogma of "penetration." This has been the biggest toad in the puddle, and has exercised an active agency in roiling and mystifying the mind of the microscopist. The doctrine of penetration as generally taken and accepted may be thus stated: objectives of low angular aperture are endowed with a peculiar inherent and intrinsic power, by virtue of which they enable the observer to see and study structures situated in different planes of the object. For example, if the objective be focussed accurately to details occupying an intermediate plane of the object to be examined, then will the low-angled glass allow the observer, without change of focus, to study other details of the said object, situated in planes either nearer or more remote. We have been taught that this is a most valuable property, and one due to the employment of low angles *only*—the idea thus conveyed being that the low angles possess a peculiar and *accommodating* power of great value to the microscopist, to which the wide apertures stand *inflexibly opposed*, and defiant.

In support of the doctrine of penetration, it has been customary to present the case of the optical principles governing the action of low apertures, contrasting the same relatively with similar conditions involved in the use of the high angles; thus we have been taught that the narrow-angled glass admits as a matter of course, but a narrow cone of light, the pencils crossing at the focal point at a very acute angle. Hence it is *"obvious"* that it matters not whether the object to be viewed be placed exactly at the crossing point or a little within or without the said focus. The accompanying and supposed increase of working distance attributable to the narrow aperture of course is not lost sight of; and we are here admonished to keep in mind the fact that, with an infinite working distance, there would be no need of special focal adjustment, and hence the longer the working distance the better. On the other hand, we are told that objectives with very high apertures admit a much wider cone of light, the lateral rays of which cross in the focal point, at a more obtuse angle, and hence the necessity of placing the object to be viewed exactly in the focal plane. On all other planes, nearer or more remote, the object being out of the crossing of the rays, cannot be well defined; and here again, conversely the presumed decrease of the working distance due to the increase of aperture is held prominently in view.

To all of the above, which has proved so acceptable to the world of microscopists, the author long ago published his dissent. He never did, and does not to-day, take the least stock in the aforesaid enunciation of the so-called doctrine of penetration.

Admitting, as in the case of the two objectives presented, that the cone of light illuminating the field from the high-angled objective is wider, and that the lateral pencils cross in the focal point at a more obtuse angle than can occur in the case of the narrow-angled glass, it is nevertheless true (and singularly this little fact seems to have been entirely lost sight of), that the wider cone of light due to the employment of the wide aperture includes *all* of the *central pencils* present in the case of the narrow-angled glass. In other words, there are just as many central pencils at work (and remember that these *are* the fellows that cross the focal plane at such an acute angle, thus furnishing the beloved pene-

tration) in the making up of the wider cone as *can*
occur with the use of the narrow aperture; furthermore,
that it would be not only possible, but eminently *practicable*, by the use of a diaphragm, to cut down the
cone of the wider aperture objective to correspond with
that of the low-angled glass; hence it is obvious that in
this latter case the two objectives would be worked
under similar conditions as respects the angle at the
crossing of the rays, and, applying the argument
based thereon, neither glass can be endowed with the
greater penetration.

Says one, "How about the working distance?"

The relations of angular aperture to working distance have already been discussed, and intentionally,
with the view of preparing the mind of the reader for
the above interrogatory. But there remain other considerations bearing on the matter of working distance,
and the clinching argument on the part of the writer
remains to be presented.

In doing this, the author is compelled to deal in
assertions dogmatically. In the handling and comparing of object-glasses, he has had a very large experience,
and he feels that he has the same liberty to speak *ex
cathedra* as has been granted to others. Moreover,
what he now has to say is "important if true," and he
is as well assured of their correctness as of any other
fact within his knowledge and experience, nor is he
alone in the matter about to be stated. Without
exception, all who have experimented in the proper
direction assent to all that will be here claimed, while

those who have not, may reasonably be expected to *know not.*

Let it be required to display an object under the microscope, and under a given amplification. It matters not what the object may be—be it a diatom, or a bit of voluntary muscle, or what not; nor does it matter as to the amplification—be it 60, 600, or 6000 diameters, as the case may be.

Now, to attack this object, we will provide two *sets* of objectives, including all the focal lengths, say from the inch, upwards, to the one-fiftieth—these glasses to be the very finest of their kind made at the present day, and notably of low apertures; the other set to be similar as to the range of focal length and quality, but notably to possess the highest apertures (respectively) known.

Now choose your object, select your amplification, and display the former, using the low angles with their very best foot foremost. This done, allow me to remove the objective, replacing in its stead *the suitable* high-angled glass, and I affirm pointedly that the object shall be equally well displayed, under the same amplification, etc., and by an objective, too, having greater working distance than the low angle first selected.

It should be contemplated, in any competitive comparisons of this kind, that they be conducted without prejudice, and solely in the interests of science, and when so conducted, and by observers fitted for the emergency, the author apprehends that his statements will be found correct.

We have thus again endeavored to make manifest that the idea that angular aperture is accompanied by a sacrifice of working distance has no real existence— that is, in the form popularly accepted.

Thus far we have discussed "angular aperture" in its popular signification, and, in several of its aspects, from the definition given from the micrographic dictionary. Taken in conjunction with the remarks we have thus far had occasion to offer, the reader would probably infer as axiomatic that the range of apertures would necessarily be confined within the axial pencil and the one striking the underside of the slide at near coincidence, thus traversing and limited by an arc measured by nearly 90°—the latter being equal to 180° of aperture. This, too, is the precise aspect to which the author desired to restrict his observations. Now there is another kind of aperture of which very little is generally known, we refer to

BALSAM APERTURES.

From a theoretical or mathematical standpoint, the study of balsam angles fairly bristles with difficulties; it has been to us a problem to which our school boy wrestlings with Euclid seem a pleasant and simple exercise. While we frankly admit our incompetency to properly present the subject, we have to remark, on the other hand that we were not willing to send forth this little book without at least some mention of the matter.

Observers interested in the history of the American

objective (and American observers ought to be, to a man) will find the subject ably discussed in the columns of the London *Microscopical Journal*, reference being made to the celebrated controversy on the subject of angular aperture between Mr. R. B. Tolles, of Boston, Mass., and Mr. Wenham of London. In this discussion, the American side of the question was ably assisted by Col. J. J. Woodward, of the U. S. Army, and Prof. Keith, of Georgetown, D. C. The entire controversy is well worth reprinting in a consolidated form, and should find an appropriate place in the library of every American observer.

No attempt will, for the reasons given, be made to discuss the subject of balsam apertures in these pages. We shall, however, try and give the novice an idea or two connected with balsam angles without which some things which will hereafter be presented, would be wholly unintelligible.

Suppose we put a ray of light *down* the tube of the microscope, thus reversing the usual order of things, and that said pencil have an angle of 41°. This pencil traversing a suitable objective in position over a balsamed mount, will find emergence into air at 90°; equal to what is recognized as 180° of aperture. Such an objective would be said to possess a *balsam* angle of 82°, in other words (rejecting fractions), the balsam angle of 82° is said to equal an air angle of 180°.

Now it is claimed by certain American opticians, that it is possible to construct immersion lenses that are capable, when worked over balsam mounts, of recognizing interior pencils greater than 41°. Mr. Tolles claims for some of his recent immersion objectives, balsam angles as high as 120°.* We have devoted a great deal of time to the study of this class of objectives.

These glasses were generally known here at home as "duplex," or four system immersions, as distinguished from the older form having a single front. Many of these objectives, ranging in their claims as to balsam angle from 82° to 100°, have passed through our hands, and have been submitted to close and careful study; one object on our part being to determine, if possible, whether a glass said to be of a low balsam angle was in any respect characteristically different from another claiming a higher balsam angle, and in this way to arrive at some determination as to the existence or validity of the claims resting on the recognition of the angle itself.

As a result obtained from close, tedious and protracted observations, dating from the present date back to that of the first "duplex" made, we unhesitatingly affirm that it is quite possible to distinguish the performance of a duplex objective of 82° balsam angle, from a similar glass of 100°, otherwise we would have had no occasion to have introduced the subject at all.

Now, in the *effect* of *balsam* aperture, we recognize in the high *balsam* angles precisely what has been attributed to high *air* apertures, namely, a decrease of working distance, as the *balsam* angle is increased, and

---

* Messrs. Tolles & Spencer are now (1880) making objectives of 120° balsam aperture.

a corresponding increase of definition, when worked by extremely oblique light over balsam mounts of "difficult tests." These characteristic differences in performance are so palpable, as to enable us to select in less than fifteen minutes' use of two glasses over the Moller probe plate, the higher-angled glass (balsam) from the lower.

It has been claimed by some who have used the duplex glasses, that the higher performance by central light is obtained with those of the smaller balsam angle. My own experience does not authorize me to endorse these conclusions. Certain it is, that when the higher angles are used by central illumination, their immense power of light will, if the matter receive not proper attention, defeat the glass, and again, even with oblique illumination, the high-angled objectives require the most careful attention and expert handling.

We are, therefore, prepared to endorse to some extent, referring to *high balsam angles*, the remarks which have been quoted from "The Microscope and Its Revelations."

FLATNESS OF FIELD.

It is, of course, desirable that an object known to be flat should so appear when viewed under an objective. The optician, however, has thus far found it impossible to secure perfection in this respect, combined with the highest aperture obtainable, and this might be urged as an objection to the use of wide-angled objectives. The slight error shown by the glasses referred to, has much more weight on paper than occurs actually in practice, where the great increase in definition obtained causes the slight deficiency in flatness of field to sink into utter insignificance. Hence, in testing the qualities of an object-glass, the flatness of its field would hardly be called into requisition.

Nevertheless, flatness of field must have its due weight, and the performance of a first-class objective should not betray any serious error. It will be well, then, when examining an objective, to look after the quality referred to.

The careful testing of an object glass in this particular, is not such an easy and off-hand matter as might at a glance be presumed; the manipulator may arrive at incorrect results, and thus condemn a glass without due cause therefor, and a word or two as to the proper mode of conducting this test may not be amiss.

First, it is of the first importance that the object be in itself flat when presented to the objective. Errors may creep in—first, because the object is not in itself flat, or, second, because it is mounted on an improper slide, or on the cover thereof—very few slides are flat, and covers are notoriously "in wind;" third, the stage of the microscope may not be at right-angles to the optical axis, and fourth, the eye-piece may not suit the objective. To guard as far as possible against these sources of error, proceed thus:

Select a fine stage micrometer, having a band of lines just about as close as the objective can well display when viewed with nearly central light, and if you have

no conveniences of your own, take this to the machinist, or the watchmaker, either of whom will allow you
the use of a steel "straight edge," with which you will
be enabled to ascertain with tolerable exactness whether
one side of the micrometer is truly a plane. This done,
by the aid of a suitable "guage," you will also determine as to the other face. If, in a process of this kind
the micrometer betrays defects, it should be discarded
and another one chosen in its place. In our own practice we always use a Nobert test-plate, which has been
found to be very reliable.

To avoid the third source of error, view the lines,
selecting as close a band of lines as possible, and using,
as before named, nearly central light, and the two-inch
eye-piece, with the micrometer placed horizontally and
vertically on the stage, examining the bands from end
to end as they appear in the field. Repeat the experiment, but *reversing* the micrometer *in each* position end
for end. Any error due to the pose of the stage is thus
made manifest.

Should you have reason to suspect trouble from the
eye-piece used, repeat the entire test on the stand of
some friend, always using the two-inch, or lowest eye-
piece. This ocular furnishing the more severe test.

Finally, let it be known that all the eminent makers,
both American and English, furnish glasses that are
not to be rejected for non-flatness of field. We would
rather trust to the reputation of these gentlemen, than
to the test conducted by the novice. A little practice
on the other hand on the part of the latter will not be
a waste of time.

MOUNTING OF OBJECTIVES.

Having glanced briefly at the optical portion, a few words in reference to the mechanism of the instrument may not be out of place.

Adjustable glasses are provided with movable or stationary fronts, that is to say, in the process of revolving the correction collar the front lens also revolves, or remains stationary, as the case may be. The stationary front being the most expensive mounting, it is generally adopted in first-class American or London glasses of short focal distance.

In the use of objectives of tolerably long focal distance, the necessity for the stationary front is not so apparent, and some first-class makers adopt either form of mounting for such objectives. It behooves the buyer to keep in mind this difference in the *cost* of the two mountings, for of two glasses, both equal in optical performance, the one adjusting with stationary front *ought* to be the most costly.

Generally, the mechanism of the collar adjustment should be first-class. There should be no "slip," "backlash," or " dead-point;" the collar should rotate with a certain firmness of action, and yet run as " smooth as oil;" there should be no undue rubbing or grating, nor " hitch or hindrance " of any kind. With a one-half inch, or a four-tenths, a slight " slip " or " back-lash " need not defeat an otherwise satisfactory objective; but, on the other hand, nothing of this sort can be allowed in the higher and first-class objectives. If the

collar be adjusted by the maker to rotate under high
pressure only, regard this with suspicion, and examine
for back-lash closely, or what is perhaps the better
way, get the opinion of a skillful mechanic.

The models of some of our American objectives are,
in the opinion of the writer, far too large and clumsy,
and the fronts too large and too flat. We decidedly
prefer the conical front — the more conical the better.
The new tenth of Mr. Herbert Spencer, which we have
before found occasion to mention, was, beyond all cavil,
the most beautifully mounted glass we had ever seen,
and in which the last-named objections were almost
wholly avoided.

NOMENCLATURE OF OBJECTIVES.

American and English microscopists usually class
their objectives on the basis of their focal length, it
being arbitrarily assumed that the inch glass worked
with ten-inch tube and with two-inch eye-piece, should
give an amplification of fifty diameters. Hence, the
half-inch glass would give one hundred diameters, the
one-fourth two hundred, the one-eighth four hundred,
the one-tenth five hundred, and so on. It very seldom
happens, however, that an object-glass will exactly respond to the designation given it by the maker — some
opticians over, while others under-rate their instruments; and then again with an adjustable glass the power
will change in different positions of the adjusting collar,
the amplification being greater with the systems at
closed than when at "open-point." Even when due

allowance has been made for this last condition, we have seen English eighths having really higher power than some American one-twelfths. Again, but a few weeks since, we handled a foreign one-sixth which was superior in amplification to a Spencer one-eighth. We, here at home, have scolded a good deal at this state of things, and at times have *rated* our English cousins for thus *underrating* their objectives. It is quite unnecessary here to traverse the ground we have already discussed, to render it evident that as the objective approaches perfection, these nominal distinctions, based on focal length, fail to have particular force. If, for example, it were possible to produce a two-inch objective, which, under extremely high eye-piecing, would more than do the work of our present tenths, then we could afford to work pretty much with one objective, and to the eye-piece look for the determination of the power. Until, however, some such "possibility" shall occur, our present nomenclature will be invested with *some* force, no matter how variable this force may be. It is, therefore, desirable sometimes to determine the actual rating of an objective.

The method usually employed, is to place a stage micrometer in position, using a ten-inch tube, and projecting the image by aid of the camera lucida on a screen ten inches distant. Knowing, then, the actual value of the divisions on the micrometer, we are enabled, by the measure of the magnified image thereof, to determine the amplification. This method is so well known and so often practiced, we content ourselves with its general mention.

The method has its drawbacks. For instance, all microscopes do not have tubes precisely ten inches in length, especially when the objective is placed in position; and then, again, if the problem be to determine which of two lenses is the stronger, it may occur, and pretty surely too, that one glass will have a shorter setting *i. e.*, "mount," than the other. There are ways to dodge the difficulty, but the best method we have seen of measuring objectives, where strict mathematical accuracy is not a vital consideration, is by the use of a formula, modified by Col. J. J. Woodward, and by him given to the public through the columns of Silliman's Journal not long ago (Vol. III., June, 1872):

Formula:

$$F = \frac{ML}{(M+1)^2}$$

where F — focal length, M — magnification *without field-glass of eye-piece*, and L — length between stage micrometer, and micrometer in eye-piece.

If the "B" (one-inch ocular with its field-lens removed, be selected, a micrometer ruled to one six-hundredth of an inch will be found to suit it very well.

It will be seen that the formula is extremely simple, and quite within the comprehension of all; we, however, append an example or two, selecting cases which have occurred in actual practice.

. (1.) We lately had in hand a glass of Mr. Gundlach's, made for a half-inch, and we desired to be assured as to its true rating. Placing a suitable micrometer on the stage, and selecting a one-inch ocular, with its field-glass removed, but fitted with micrometer ruled to one six-hundredths, and having brought the lines of the stage micrometer in focus, we then measured the distance from micrometer to micrometer, finding the same to be just ten inches, and on comparing the divisions of the stage micrometer with those of the eye-piece, we found those of the stage micrometer to be amplified eighteen linear. Now, working out the formula, and substituting for the symbols their proper values, we have

$$\text{Focal length} = \frac{18+10}{19^2} = \frac{180}{361} = \cdot 5 \text{ in. or } \frac{1}{2} \text{ inch.}$$

Hence, we find the glass exactly to the rate given it by its maker.

(2.) In a similar manner it was desirable to test a (so-called) one-fifth objective. Proceeding just as before, simply changing objectives, we measured again the distance from micrometer to micrometer, getting again just ten inches—a matter of chance, however, and one not likely often to occur. Comparing the two micrometers again, we get an amplification of the stage lines of thirty-eight linear. Substituting values as before we now have

$$\text{Focal length} = \frac{38+10}{39^2} = \frac{380}{1521} = .25 \text{ in. or } \frac{1}{4} \text{ inch.}$$

For the purpose of developing the capacity and value of the formula, we will now repeat this example, but in place of using a distance of ten inches, we will arrange to secure in lieu thereof a distance of twelve and one-half inches. We now find the amplification, as shown by the two micrometers at this greater distance, to be forty-eight linear, and substituting this value, we have

$$\text{Focal length} = \frac{48+12\cdot50}{49^2} = \frac{600}{2401} = \cdot25 \text{ in. again.}$$

Either computation gives the same results (rejecting fractions too small to affect things sensibly), and we see, too, that the glass sold for the one-fifth is really but a true fourth; and, furthermore, it is evident that a change in the distance of the two micrometers — in other words, working with different lengths of tubes— does not practically affect the results.

TABLE OF THE MAGNIFYING POWERS OF SINGLE CONVEX LENSES.

The first column gives the focal length for parallel rays; the second, third, fourth and fifth, give the magnifying powers at ten, twelve and a half, twenty-five, and fifty inches respectively, the last three being from the *American Journal of Science and Arts*, as before mentioned, while the second was computed by the Hon. J. D. Cox, of Ohio, and by him presented to the author.

Focal Length for Parallel Rays.	Magnifying Powers at 10 inches.	Magnifying Powers at 10¼ inches.	Magnifying Powers at 25 inches.	Magnifying Powers at 50 inches.
1	2	3	4	5
3 inches.	1·33¼	1·50	6·17	14·59
2 "	3·00	4·00	10·40	22·95

1 1-2 "	4·67	6 17	14·59	31·30
1 "	8·00	10·40	22·95	47·99
2-3 "	13·00	16·69	85·47	72·98
1-2 "	18·00	....	....	....
4-10 "	23·00	29·21	60·48	122·99
1-4 "	38·00	47·99	97·98	197·99
1-5 "	48·00	60·48	122·99	247·99
1-6 "	58·00	72·98	147·99	297·99
1-8 "	78·00	97·98	197·99	397.99
1-10 "	98·00 .	122.99	247·99	497.99
1-12 "	118.00	147·99	297.39	597.99
1-15 "	148·00	185.49	372·99	741·99
1-16 "	158·00	197.99	397·99	797.99
1-18 "	178·00	222·99	447·99	897·99
1-20 "	198·00	247·99	497·99	997·99
1-25 "	248·00	310.49	622·99	1247·99
1-50 "	498·00	622·99	1247·99	2497·99

By the use of the preceding table one is able to measure the focal length of any objective, and without involving the trouble of computations. In ordinary cases
it will be only necessary to see that the two micrometers shall be exactly ten or twelve and one-half inches
apart. The amplification in either case being noted, we
enter the table under the proper head, and corresponding to the amplification will be found the focal distance.
For instance, taking the first case we have presented,

that of Mr. Gundlach's objective, we worked it a distance of ten inches, and observed an amplification of eighteen linear. Entering column two of the table, we find at a glance eighteen to be opposite to focal length of one-half inches.

In the case of the so-called fifth, with the two micrometers at ten inches distance, we observed a magnification of thirty-eight linear; entering the second column of the table again, opposite thirty-eight, we have the focal length, equal one-fourth inch. To the last computation we worked the same case, but with the micrometers twelve and one-half inches distant; in this case observing a magnification of 48. Entering the table in column third, opposite to 47.99, we have the same focal length as before, viz., one-fourth inch.

It often occurs that many first-class stands cannot be worked at a distance of ten inches with objectives of ordinary lengths. In this case use the twelve and one-half inch distance, which can always be obtained by using the draw tube.

CHAPTER III.

What constitutes a really superlative wide-angled objective? It has been our fortune to reply to this question upwards of ten thousand times, and it may be true that no two of these responses have been exactly alike. In the present essay let us get close to the reader, introducing a little gossip, and, at the same time, seeking a little relief by avoiding expressions which are getting to be stereotyped, such as the "writer," "this," "the author," "that," etc., helping ourselves freely to the first person singular, about as one would when writing to a friend.

It has been already shown that there are two classes of high-angled glasses—namely, those having balsam apertures, say up to 100°, the other class responding to *air* angles, up, say, to the "impossible," 180°. And, first, let us consider the objective of high balsam angle.

I have, in advance, stated that these glasses necessarily have a very short working distance. But this remark must not be swallowed whole; let it be taken rather with a "pinch of salt," keeping this little fact *well* in hand, that there are wide-angled balsam apertures, having *greater* working distance than will be found present in other objectives of the same nominal focal length, and having, too, no really wide air angles.

In comparison with such as these, the high balsam apertures have nothing to fear. As to working distance, a vast amount of misguided effort has been expended in the vain attempt of trying to compare one kind of an objective with another; for instance, a wide-angled air aperture with objectives of narrow air angles. Hence, we have had nothing but muddle and confusion. I repeat what I have before printed, that any attempt of the kind is as futile in its very nature as it would be to compare a " turnip with an orange." It has, therefore, been my aim, however imperfectly I may have succeeded, to lead my readers along one road at a time, and this " objective " point has been steadily held in view while writing this little book, and up to the present writing, whether success has attended the effort or not, I claim some credit for having essayed in this direction.

" Revenons a nos moutons."—I repeat, then, that of two superlative objectives having balsam angles ranging from 82° to 100°, the rule presented holds good, i. e., that the higher aperture will have the shorter working distance, or, in other parlance, the gain in aperture will be accompanied by a sacrifice of working distance.

Now, we all know what is lost with the decrease of working distance. Let us, therefore, seek as to what is gained by the increase of aperture. First, we gain a wonderful increase in intensity of *definition;* an increase in definition too, entirely unapproachable from any other direction. In this particular these glasses stand alone and defiant. Secondly, their immense power of light, superlatively corrected as they are, enable them to be used under wonderfully high eye-piecing, and with comparatively slight loss of either light or definition, while by the aid of these high oculars, the greatest magnifications are obtainable, and in these particulars also, do these objectives again stand alone and defiant. ' Somehow, among the few who have paid attention to the claims of the balsam apertures, it has got to be the popular impression that it is the particular province of these objectives to bear high oculars, the impression having force in a restricted, or limited sense — to wit, that there is nothing gained really by the use of high balsam angles under low or medium oculars.

Any idea of this sort is totally in error, for even under a low or medium eye-piece, the higher balsam angle will demonstrate its presence by an increase in the intensity of definition, while its greater power of light is in *itself* a power capable of being turned to certain necessary and useful purposes. I have already stated that it is quite possible to distinguish the one glass from the other, and, I might have added, " under low or medium oculars."

But, as I have said, with all this gain, there is an accompanying loss of working distance. But this is not all; those who have been accustomed to work all their lives with objectives of low or medium apertures, might with good reason, claim that the high balsam angles are " exceedingly inconvenient to use." Palpable is the fact, that the slightest error in the collar adjustment, or in the management of the illumination, will suffice to defeat the maximum performance of the balsam apertures, and the higher the balsam angle, the greater the " bother " attending their use.

Nor is there anything so exceptional in all this. In any instrument of precision, when, by successive improvements, we gain width of effective range, and, simultaneously, a nearer approach to accuracy of determination, then, in ninety-five cases out of the hundred, do we, in like manner, increase the complications of the instrument; thus introducing " inconvenience " and " bother," and, be it known, *an accompanying call for skilled manipulators.*

We often hear another remark — to wit, that those owning and using the class of objectives we are now considering, do nothing else than " fight them." Before attaching any definite force to this remark, it will be well to inquire, " for what purpose is the battle!" Is it to determine which of two objectives, differing in construction, but supposed to be nearly alike in performance, is the better one, and with the purpose (avowed or otherwise) of rendering some contribution to a scientific end; or is it that " A " battles " B," the object being on the part of " A " to assure himself that his glasses are not behind the age in quality of performance; or, again, may it be that " C," being requested by his inexperienced friend " D " to select for him an objective in every respect fully up to the times? Or, to mention a very possible case, does " E," who, having a few hours of daily leisure, desires to appropriate the same to the study of object-glasses, with the view of perfecting himself in the manipulations of the same, that he may in turn render his observations and experience of value to those who devote their spare time to investigations with the instrument? If these *are* the ends in view to be accomplished, then I say, " fight objectives," and furthermore that " E " need not be ashamed of his occupation. On the other hand, *if* objectives are fought for other and baser purposes, that is no affair of mine. I cannot help it, and the *objective* is not to blame.

I claim, then, that the high balsam angles are indispensable in the studies of the advanced investigator; they alone, " bother or no bother," are the ones for the work. Let the reader keep constantly in mind that, in the study of difficult and delicate structures, the slightest superiority of definition is of vital importance. A first-class objective of high balsam angle will show, clearly and accurately, details in an object, which, from their extreme tenuity or transparency, would be totally invisible when viewed under an objective having but air angle.

To sum up: We thus arrive at the conclusion that objectives of high balsam angles, have, as compared with others of less balsam angle, a shorter working distance, and in general are " inconvenient and bothersome to use; nevertheless, that their use cannot be dispensed with in the study of delicate and difficult structures, and it may be further stated that these are the glasses suitable for the study of difficult diatoms, the display of the Nobert nineteenth band, and the like.

Thus far, whatever force may attach to the elements of "inconvenience" and "bother," of which we have heard so much, I have allowed full play and weight; but let the reader not lose sight of *the fact* that those accustomed to the use of adjusting glasses never abandon them to return to their first love, nor do they regard the process of adjusting a superior objective as being in any way or shape a "bother" or "inconvenience." It may be well, right here, to state a bit of personal experience. I am in the habit of putting adjusting glasses in the hands of my pupils at the earliest possible moment; not that they will accomplish more with them for a time than would be the case with those non-adjustable, but in order that the pupil be early brought in contact with them, and thus, by degrees, get accustomed to their use. By adopting this course, the tyro learns, in due and proper season, an important lesson, to wit, that there are some things to be done besides putting an object at one end of the tube and the eye to the other in order to see what's what. Now, it often happens that there may be more pupils in the laboratory than adjustable objectives, and thus, perforce, a student or two will have to fall back on the non-adjustable. When this state of things occur there is inevitably indications of discontent, and it is never on the part of those pupils occupied with the adjustable objectives.

So much for the "inconvenience" and "bother."

Now, let us get after that other elephant — working distance. I have granted that with the higher balsam

apertures there is a loss of distance. Now I propose
to discuss this " loss " from a practical point, and in so
doing shall state facts, including names and dates.

On the 22d of March, 1878, I received from Messrs.
C. A. Spencer & Sons, a one-tenth objective of high
balsam angle. This glass was made to my order, and
for the Hon. J. D. Cox, of Ohio. Before sending it to
its destination I " fought it." Its performance was as
follows: Worked over the balsamed Moller probe.
plate, on the Zentmayer histological stand, with the
mirror-bar swung to 31° from axis, I saw distinctly the
striæ of Nos. 18 and 19 of said plate, nor had I any
doubt as to thus seeing the No. 20, the illumination
being a common small coal-oil lamp. The light was
taken direct from the mirror.

Worked over the same Moller plate, but with the
Wenham reflex illuminator (an instrument devised for
shutting out all pencils having air transmission), I
saw the transverse striæ of the No. 20 of said plate
handsomely using powers from 1,000 to 4,000 diameters.

Tested over a fine slide of English podura, using
ordinary illumination, and nearly central, I had the
finest display I had ever witnessed.

Worked over *navicula angulata*, with "dead central "
illumination, it gave me instantly the markings very
handsomely.

Worked over a dry mount of *amphipleura pellucida*.
These shells surrendered at discretion ; ordinary oblique
illumination.   Substituting a balsam slide of this

diatom, placing the same in position *under* the stage of the histological, I got handsome resolutions with the mirror, at say 80° from axis, and also when the swinging-bar was posed so that more than half the mirror was above the stage.

Tested for working distance. I found the same to be .015, equal, say, to one-sixtieth of an inch.

There were very many other "rounds" to this fight, which are purposely omitted as having no bearing on the subject in hand. The reader has before him the work of the objective by central, centrally disposed, tolerably oblique and *decidedly* oblique illumination.

Now, thus having the work of the objective before us, it is palpably evident that it has all desirable working distance. Surely no one but a bungler would attempt to cover a nice mount with glass over one one-hundredth of an inch in thickness, while, with such a cover, the one-tenth will have plenty of room and to spare; and if we compare the glass named with the usual run of immersion tenths, it will be obvious that there is no loss of working distance nor any sacrifice thereof in any way, shape, or manner.

OBJECTIVES OF LOWER BALSAM ANGLE.

These, when compared with the glasses we have just had under consideration, will be found to have greater working distance, as has been before stated; and with this increase of distance there will be greater focal depth — the so-called "penetration" of the books. Now, in the preliminary investigation of many objects,

this is a desirable quality in a glass.   It enables us to
search through an extemporized mount in the least possible time.   And then, again, we are less liable to allow
important details of structure to escape our attention.
These *are* advantages, and must be recognized as such;
but, per contra, let it be remembered that in subsequent examinations it often becomes quite as important
that this quality of penetration should be absent.   For
instance, when it is desired to study structures situated
in one plane, and one plane *only*, the less of penetration the better.   I have not space to enlarge on this,
but let the reader not forget the fact.   The glasses of
the lower balsam apertures are really the easiest to
manage, and yet are effective and adequate for a large
class of work.   They have less power of light, and
hence do not "stand up" so well under high eye-
piecing, nor have they the same exquisite intensity of
definition.   In a certain sense, the diminution of definition is very slight, indeed scarcely to be noticed (if at
all) when examining tolerably vigorous tests.   Hence,
they are quite adequate for the resolution of nearly all
the recognized test objects when mounted *dry*.   It is,
then, over exceedingly thin, faint, delicate, and transparent tests that the higher angles assert their suprem-
acy.

If the general definition as to what constitutes a
high-angled objective, given on a preceding page, be
accepted, then there remains room to discuss a class of
objectives of such focal lengths, nominally, as defy any
expert effort on the part of the optician to extend

10 Microscopy.

their aperture to any thing like an approach to 180°.
Among such may be included the two inch of 25°, the
one and one-half inch of 37°, the one inch of 40°, the
two-thirds of 45°, the one-half inch of 100°, the four-
tenths of 150°, and so on. All the glasses above
named may be considered as high-angled objectives.
They should be nicely corrected, fully up to the limits
of the aperture claimed for them, and, as compared
with similar lenses of narrow apertures, they should
possess a much greater power of light, and bear higher
eye-piecing; and, let it be borne in mind, that of the
above-named objectives, the one having the greater distance will be "endowed" with the greater "penetration."

And right at this place let us see what the "power
to bear high eye-piecing" means. To illustrate this,
we will take a case such as might happen in practice.
For example, I desire to see the transverse lines of
pleurosigma balticum, and, as mounted on the Moller
balsam test plate, their striations being from 31 to 34,
in .001 Eng. inch. Now, I attack this with, say, a
real good low-angled one-half inch, and find that I can
just get a glimpse of the lines with the one-half inch
eye-piece. I then apply the one-quarter inch eye-piece,
and find that the view is not nearly as satisfactory as it
was with the lower ocular; hence I conclude that the
one-half inch eye-piece is as high as the objective will
bear. Now, removing the one-half inch objective, we
will substitute a two-thirds, but of higher aperture.
With the latter glass, using the one-half inch ocular, I

get better shows than was the case with the former objective, and on applying the one-quarter eye-piece I
have better definition yet. There is some loss of light,
to be sure, but there is enough left to do the work, and
in a more satisfactory manner than occurred when the
lower eye-piece was in position. I take this case as an
illustrative one, because I have used it repeatedly with
my visitors, and the demonstration has, 1 believe,
always been accepted as satisfactory. It will be noticed
that thus using the two-thirds, the power applied at
the eye-piece was greater than the rating of the objective employed, and a tolerable test for an object-glass of
such low nominal power. We have here, then, in the
experiment cited, an illustration of what may be said
to be " power to bear high eye-piecing."

There is a general and indefinite idea afloat, that
there is something about high eye-piecing which ought
to be condemned. Without making any special attack,
it may be admitted that observers "came honestly by
it." The facts warrant this much: power — *i. e.*, amplification or magnification — when not wanted, is to be
condemned, and it matters not how obtained, whether
at one end of the tube or the other. On the contrary,
if any thing valuable is to be obtained by amplification,
then we are authorized to use it, and to the best advantage apply it at either end of the tube, as the case may
warrant. There seems no more propriety in the con-
demnation of high eye-piecing *per se* than ought to obtain in the use of objectives of short focal length.

ADJUSTABLE OBJECTIVES.

The generally received notion in regard to these glasses is, that they are provided with an adjusting collar, so that they may be "corrected for aberrations," due to the thickness of cover employed on the mount under observation. This is all true in a certain sense, and to a certain extent only. If we examine the same object under the same objective (a first-class glass of high aperture), but through covers of varying thickness, we will find that the very best performance of the glass will be obtained when worked over a particular thickness of cover, and that any change from this *particular thickness* will interfere with the performance of the lens, no matter how perfect the corrections in each case may be made. My own experience teaches me that the maximum definition is only obtained when the objective adjusts at or near that point in the collar adjustment which corresponds to the maximum aperture of the objective. Now, in most glasses of high angles the maximum angle will be found at only one position of the collar. I say "most objectives," for we have a one-tenth made by Mr. Tolles, that has maximum angle nearly at any point within the range of its collar adjustment. We have also a one-sixth by the same maker, which has maximum angle only at the "closed point," the aperture decreasing rapidly as the collar approaches "open," and *with the decrease of angle there is an accompanying decrease of definition and effective force of the objective.*

Here we have an argument in *favor of wide angles* which has been in the past quite lost sight of, and was for the first time presented to the public by the writer in one of his monthly contributions years ago; but all this is foreign to our present purpose.

Now, in the case of the one-sixth of Mr. Tolles, should it be attempted to use this glass over the thin cover of the Moller probbe plate, the result would only be to defeat the performance, for the glass would "correct" within three divisions of " open-point." But, per contrary, if we desire to secure the best definition of the said objective, and over the said plate, it becomes necessary to supplement by thicker cover (making sure of optical contact by the use of water or glycerine), until the glass shall " correct" at or near " closed."

Hence, it occurs that it is essential that the observer who proposes to employ objectives of wide apertures should pay some attention to the condition of things; that he should know precisely over what covers he will be enabled to get the maximum performance. In short, to ascertain where he may find the maximum aperture of his objective.

And to this end, we propose to aid and assist, as far as possible, in some future remarks we shall have occasion to make regarding the manipulations of the objective. During the past twelve years the author has received hundreds of letters from as many individuals commencing the study of the microscope, desiring such information as to the best way of making an investment as may pertain to the selection of stand, objectives, etc. The said letters have met with a prompt response, and, as has before been stated, perhaps no two of these have been alike. In each and every case there has been some dissimilarity of circumstance, or, on the other hand, we have suffered some change in our own views. Be all this as it may, the experience of the last two years enables us to speak more precisely to the point than before, and we now endeavor to respond to the interrogatory—" What shall I procure for an outfit?"

There can be no general rule that will apply to all. Let us take the following as a typical letter for consideration:

"Dear Sir: I have read some of your contributions to the *American Journal of Microscopy*. . . . I am a physician of ten years' practice; am located in a town of some ten thousand inhabitants. I am satisfied that I ought to know enough about microscopy to enable me to examine intelligently urinary deposits, cancerous growths, etc., and to this end do I propose to purchase the necessary equipment. Any information that you may be pleased to give me, will be with pleasure received," etc.

Now, I would answer this letter, and did answer it, thus: When you buy a stand get one that you will have no occasion to sell at a ruinous sacrifice. I recommend that you purchase one of the cheap and moderate-priced instruments, and, at the same time, one that will do any and all work. Such a stand ought not to cost, with one eye-piece, more than forty-five dollars (and such are described in this little book). Now, as to objectives. You can do all the work named with a one-inch of tolerably low angle, costing you, say, some $7.00; and a real good three-tenths of 70° aperture, which will cost you, say, $11.00. The chances are, however, that as you become familiar with the use of the instrument, thereby learning its value to you in your daily professional practice, you will feel an inclination to dip somewhat deeper into the problems which will most assuredly surround you. In your examinations of urines, you will get glimpses of bacteria; you may meet with structures which you are almost assured are "pale hyaline" tube casts; and you naturally desire a little more amplification and definition thereon to enable you to pronounce with certainty. In fact, you now want such a glass as a dry one-fourth of 110° or 120°, and adjustable. Now, to use this glass to advantage, you have first to become familiar with its manipulations. It will be requisite that you arrive at some knowledge of this before such an objective can be of much avail to you. Now, if you purchase the one-fourth recommended, you will still have room for the employment of the three-tenths, but you will have to study the use of the one-fourth just the same as if you had never seen an object-glass. Hence, I had rather recommend that you purchase the one-fourth at the start, and thus get *early* accustomed to the use of adjustable glasses; and in this latter case it will almost assuredly occur that you will eventually desire to employ glasses of the widest apertures, and the experience you have gained with the use of the one-fourth will be of the utmost value to you. Moreover, the one-fourth, too, will continue to be a useful intermediate glass.

And thus in replying to all my correspondents, I recognize the importance that one and all shall early become acquainted with the manipulations of adjusting glasses. To accomplish this they must use objectives of a reliable class, *i. e.*, such as will respond promptly to change of collar-adjustment, keeping well in mind the importance of buying nothing to be discarded or thrown out of use in the future. In the case under consideration, it will happen, in nine cases out of ten, that in less than two years the buyer will feel that he needs a first-class inch, or perhaps a two-thirds, in which event the old inch will be of great service as a sub-stage condenser, providing that the stand has been selected with this end in view.

It may further happen, in truth it will be *likely* to happen, that in the course of one or two years, my correspondent will either push his own observations or desire to keep pace with those of others, and over structures of the most delicate and "difficult" character, and now he will need a one-sixth or a one-tenth of the widest possible aperture. Allowing this to occur he will have expended but $110, which is less than the usual cost of a nominally first-class one-sixteenth, and all the glasses on hand will still be of service. Besides all this, he is well armed and equipped for any work requiring powers from 50 to 5,000 diameters, and there will be no objectives on hand that will not be worth their cost.

The above programme seems to me to fill the bill as well as any other that could be named, and is prac- tically the same as that followed in my own practice, having daily a large amount of work over urinary de- posits, a considerable portion of which does not require excessively fine definition, and time being very much of an object, I employ, in addition to the glasses named, a cheap, but good and reliable half-inch, say of 35° or 40°. With this glass I am enabled to perform much prelimi- nary work over liquids, and am able to dispense with covers, and the objective, from its long working dis- tance, is out of the fumes of the re-agents constantly in use. The half-inch is thus of great service, as a time-saver — in short, as a convenience.

Finally, let it be understood that I have no war to make on objectives of medium apertures. We have stated our experience, and it must pass for what it is worth. The principal point which I desire to impress on the mind of the reader, is that, for the higher class of investigations, the objectives of wide apertures stand alone, unapproachable and unexcelled; that they alone are the instruments for such work, and this, too, regard- less of the character of the illumination employed, be it central or oblique. Another point is that the use of the high angle enables the observer to cut down the numerical force of the glasses employed, thus saving unnecessary expense. We wish also to render the fact obvious, that a wide angle glass requires time, study, and attention, and that the manipulations of the same are not acquired in a day.

In my intercourse with microscopists, I have nearly always found one idea salient. The general question has been and now is, " How shall we contrive to save expense; where can we buy the cheapest? "

It seems to me that these considerations must eventually force the wide apertures into use. I was about to say, " There's millions in it." For example, we know of a microscopist who has, by purchase, from time to time, acquired a battery of some 35 or 40 objectives. Now, allowing the cost of these to average $10 each — not a high estimate — the owner has thus paid out some $350 or $400 for objectives alone! While one-half that money, expended in glasses of the highest apertures, would have enabled our friend to have accomplished any and all work that *can* be effected with his entire battery. And unless I am greatly mistaken, the wide apertures would carry the day, and with flying colors.

This is no solitary case. We have in very many instances been called on to select wide-angled objectives for our friends and correspondents, and in each and every case—there have been no exceptions—we have received assurances from the parties interested, that there would have been money saved had the facts been known in time. Says one: " My one-fifteenth that I bought two years ago is a dead letter; what shall I do with it? I can't sell it for anything *like* what it cost, it would be perpetrating a swindle!" Says another: " If I had met you thirty days ago I would be a hundred dollars better off than I am to-day; you have made me sick of my objectives." I say, therefore, that it seems to me that this string *must* pull, to the end that the microscopist, who desires to invest his money to the best advantage, will be driven to the use of the wide apertures, no matter how strongly he may have become prejudiced in another direction. We believe and earnestly hope, that in presenting the subject to the readers of this little work, we have done them a good honest turn, and one that will be appreciated.

Before entering on another part of the subject of this book, we desire to add that the views thus set forth are the result of practical experience. Here is no speculation; but, on the other hand, the reader has before him the fruit of protracted and close study. The time was when we were almost alone in our opinions, but that date is numbered with the past. Among those who had read his printed articles in their various periodical forms, and have subsequently visited him, "to see for themselves," he can name many of his dearest friends, all of whom have, in turn, placed the writer under countless obligations. And, right here, by the way, let me say, that, as an interesting and enjoyable pastime, that of "fighting objectives" is, when occasionally indulged in, second to no "2.25" race on record! But of any of the good things of this world—"enough is a feast."

EYE-PIECES.

It would be a real improvement to assign to individual eye-pieces their actual power expressed in inches, and to dispense with the arbitrary and indefinite nomenclature formerly, and, to a great extent, at present in vogue. Thus, let the two-inch have the same magnifying power as a simple lens of two inches focus, and similarly as to the other oculars used. Messrs. Spencer, Tolles, Sidle, and we believe Zentmayer, adhere strictly to this plan.

We use principally the inch, half inch, and one-fourth inch. For those oculars above the inch in power, we greatly prefer the solid eye-pieces. We have also the two-inch and the solid one-eighth inch. The former is however, but very seldom used; in fact, in our own practice, it is seldom wanted or even thought of. The solid one-eighth, although not in general use, cannot be dispensed with when working with sunlight illumination in conjunction with objectives of high balsam angles. We cannot too strongly recommend that every attention be bestowed as to the quality of the solid oculars, and, as a rule, that they be made expressly to order.

It occasionally happens that stands fitted with poor objectives are placed in the market for sale, while to accommodate the quality of the objectives the powers of the oculars are let down considerably—thus illustrating the truth of the well-known proverb, *c'est le premier pas qui coute.*

Let the purchaser look well to these points before consummating an investment.

CHAPTER IV.

This instrument, and its various modifications, having come into very general use in this country, we give the following cut and description, taken from Dr. Carpenter's work on " The Microscope and Its Revelations:"

"A very ingenious and valuable illuminator for high powers has been recently devised by Mr. Wenham, and constructed by Messrs. Ross. It is composed of a glass cylinder, half an inch long and four-tenths in diameter, the lower convex surface of which is polished to a radius of four-tenths. The top is flat and polished. Starting from the bottom edge, the cylinder is worked off to a polished face at an angle of 64°; close beneath the cylinder is set a plano-convex lens of 1¼-inch focus. When parallel rays are thrown up through this apparatus from the mirror, they impinge on the upper surface of a glass slide at an angle of total reflection; but if a suitable object adhere to that surface, the light reaches it on an angle that admits its passage. The object is then seen brilliantly lit up upon a dark ground, and many fine markings, that escape notice with other methods, become very distinct. It is advisable to rotate the apparatus until the best position is obtained. Some skill and practice are required to use this apparatus to advantage, but it will amply repay the trouble of mastering its

difficulties. It is best suited to thin and flat objects;
with those that are thick and irregular, distortion is
unavoidable. Although specially designed as a dark

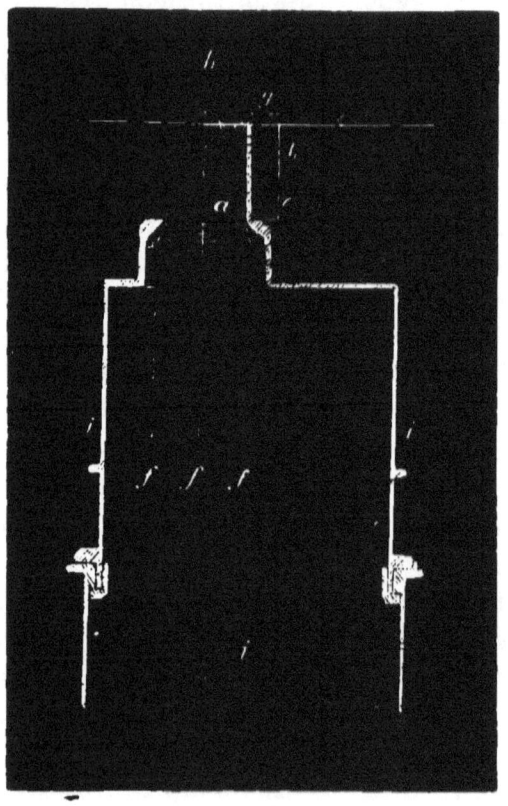

ground illuminator, good effects can with care [and by
the use of suitable objectives], be obtained for such objects as difficult diatoms, in balsam or dammar. *But
the effect is that of very oblique transparent illumination.*"

We have preferred to give the above description from

Dr. Carpenter. The interpolation within the brackets, and the italics are our own.

As is above set forth, this instrument was designed by its maker for a dark ground illuminator, and for use with high powers; was made and sold for this purpose, and was generally so used. Shortly after their introduction into this country, Mr. Samuel Wells, of Boston, Mass., made the discovery that the "reflex" when worked with certain very high-angled objectives, could be made to give a brilliantly-lighted field, accompanied with beautiful resolutions of the severest tests, when the latter were mounted in balsam, the effect being, as Dr. Carpenter most justly observes, "*that of very oblique illumination.*"

As is evident from an inspection of the cut, the instrument can be variously modified in its working, simply by changing the angle of the facet. The fact, too, appealed almost simultaneously to all, that a serious objection to the instrument was that it could be only used over balsam mounts, and of course with objectives of the highest angles. The "reflex" was therefore modified by the London opticians to suit the angles of their leading objectives, and, at the same time, or at least shortly after, similar modifications of the instrument were made here. There has been an effort here at home to claim, on the part of our opticians, some originality of design in the construction of these various modifications. The entire plan, however, belongs to Mr. Wenham, and the simple changing of the angle of the facet does not authorize any claim of invention   The author has used several of these instruments; the first one was made by Mr. Tolles, the angle of facet being 26° Another one made some little time afterwards had facet of 18°. Subsequently Messrs. Spencer have made him several having angle of 15°. The genuine instrument, as made by Mr. Wenham, is only suitable for use with objectives of high balsam angle, while those of the modern form can be employed with more or less success in conjunction with glasses possessed of tolerably wide air apertures. We have spent a great deal of time in the study of this instrument, including the several patterns named, and we find that when used with glasses of the highest balsam apertures, such as are made by Messrs. Spencer or Mr. Tolles, and over the severest tests, such as *amphipleura pellucida*, the resolutions are quite as strong and satisfactory when the illumination is obtained from the modified " reflex" as when the instrument described by Dr. Carpenter is selected, while the latter has the serious disadvantage of being adapted for use over balsam mounts only.

The instrument can be adapted to almost any good and reliable stand, even should the latter not be provided with centring apparatus; yet if the sub-stage collar be itself well centered, a little ingenuity on the part of the observer will secure good results. It is, however, a great convenience, when using the " reflex," to employ a stand allowing the sub-stage to approach the slide or to be drawn away therefrom.

To use the " reflex," place it in the sub-stage fitting, select a low power, say a half-inch or a two-thirds, and one that you know centres well with the objective to be employed. With such a glass see if the central mark placed on the instrument by its maker occupies the centre of the field. If but slightly out of centre, a very slight tipping of the instrument in its sub-stage fitting will often be all that is necessary. If the stand has centring screws these will come well into play. We have found it desirable to secure perfect centring, and at the same time make sure that the plane face of the facet is parallel to the right-hand edge of the stage, and adjacent thereto. Having placed a small drop of gly-cerine (which is preferable to water, as it does not dry) on the top of the facet, and having placed the object and slide in position, the " reflex " is made to approach and the glycerine to contact the slide. It is better not to allow the top face of the instrument to actually touch the under surface of the slide, still keeping it, however, so close thereto that the drop of glycerine shall be considerably flattened.

We are now ready to screw on the object-glass, and to make immersion contact with the cover. For illumina-tion we need one of the lowest and smallest kerosene hand-lamps that can be found. If the stand is small and low, it will become necessary to place it on a suit-able box, the tube inclined to a convenient angle. Re-moving the mirror, use the light direct from the lamp keeping the latter for the present in nearly or quite a central position. If, on looking through the eye-piece, there be found a lack of illumination, the trouble can be rectified by moving the lamp slightly. Having

11 Microscopy.

thus got light enough, next proceed to find your object and to focus the same, and if the approximate correction for the objective is known it will be well now to apply it.

Next, we proceed to attempt the proper display of the object. Seize the lamp with the right hand, turn the edge of the flame edgewise to the illuminator, and move the lamp bodily, sliding it on the table in all directions, to the right, to the left, towards the stand away from it. If, while thus manipulating the lamp, the field should be lighted with a succession of colors, blue, red, green, etc., you may be pretty sure that things are working well. Making sure, now, that the edge of the lamp flame is turned to the " reflex," and still grasping the lamp with the right hand, and the fine adjustment with the left, try the effect of very small movements of the lamp; these will be followed with a change of color in the field. Supposing, now, that the distance of the lamp from the stand is just right. We will find that by pushing it horizontally and parallel to the front edge of the stage, towards the left, we shall shortly have a blue field, while with a little further shove in the same direction we lose our illumination. Now it is in such positions, and just at the point where the illumination begins to decrease, that I get the strongest resolutions of severe tests. If we now retrace the path of the lamp, carrying the same towards the right, the field will become tinged with red, and by pushing the lamp still further thereto this will deepen, while another movement of the lamp in the same direction will cause the illumination to die away. Now, again, good resolutions can be obtained in the red similarly to those in the blue, *i. e.*, just at the point where the illumination begins to decrease. It will now be quite time for those who have never before used the instrument to look carefully to the correction of the objective, and for the trial experiment the object selected should be one that the observer can master by the usual oblique illumination. It only remains now to assure curselves that the lamp is at the proper distance from the stand, which is accomplished by simply moving it farther away, or, *per contra*, bringing it nearer, sliding from right to left, and trying the red and blue fields, as before instructed. The very best position of the lamp as to distance is generally attested by the general superior brilliancy of the illumination, together with the fact that when this distance is just right there will be room for greater lateral play of the radiant without losing the illumination. Care should be constantly observed to keep the lamp flame exactly edgewise. It may be further stated that when things are nearly in proper position, the smallest imaginable movement of the lamp will often produce marked effects.

THE WOODWARD ILLUMINATOR.

For this novel and useful accessory microscopists are indebted to Col. J. J. Woodward, of the U. S. Army. A paper, giving a detailed account of the instrument, by Col. Woodward, was read before the London Royal Microscopical Society, and subsequently published in the London *Monthly Microscopical Journal*. The paper is written with singular tact and perspicuity, and is here reprinted unchanged:

A Simple Device for the Illumination of Balsam-mounted Objects for Examination with certain, Immersion Objectives whose "Balsam Angle" is 90° or upwards. By Surgeon J. J. WOODWARD, Brevet Lieut. Col. U. S. Army.

[Taken as read before the ROYAL MICROSCOPICAL SOCIETY, June 6, 1877.]

" Certain immersion objectives are so constructed that they are capable of admitting rays which enter the front lens at a greater angle with the optical axis than the limit for dry objectives.   That this is not only theoretically possible, but that such objectives have been successfully constructed, was several years since demonstrated in the " *Monthly Microscopical Journal,*" both by Mr. Keith and myself,* notwithstanding which the contrary has often since been energetically asserted by writers in the same Journal.

" Meanwhile, immersion lenses possessed of the excessive angle in dispute, continue to be put into the market by more than one maker; and perhaps some of the purchasers will be interested in a simple device which I have used for some time with such objectives to illuminate test-objects mounted in balsam.   This device consists merely of a right-angled prism of crown glass mounted beneath the stage in such a manner that its long side can be connected by oil of cloves, or some similar fluid,

* June, 1873, p. 268; November, 1873, p. 210. March, 1874, p. 119; September, 1876, p. 124.

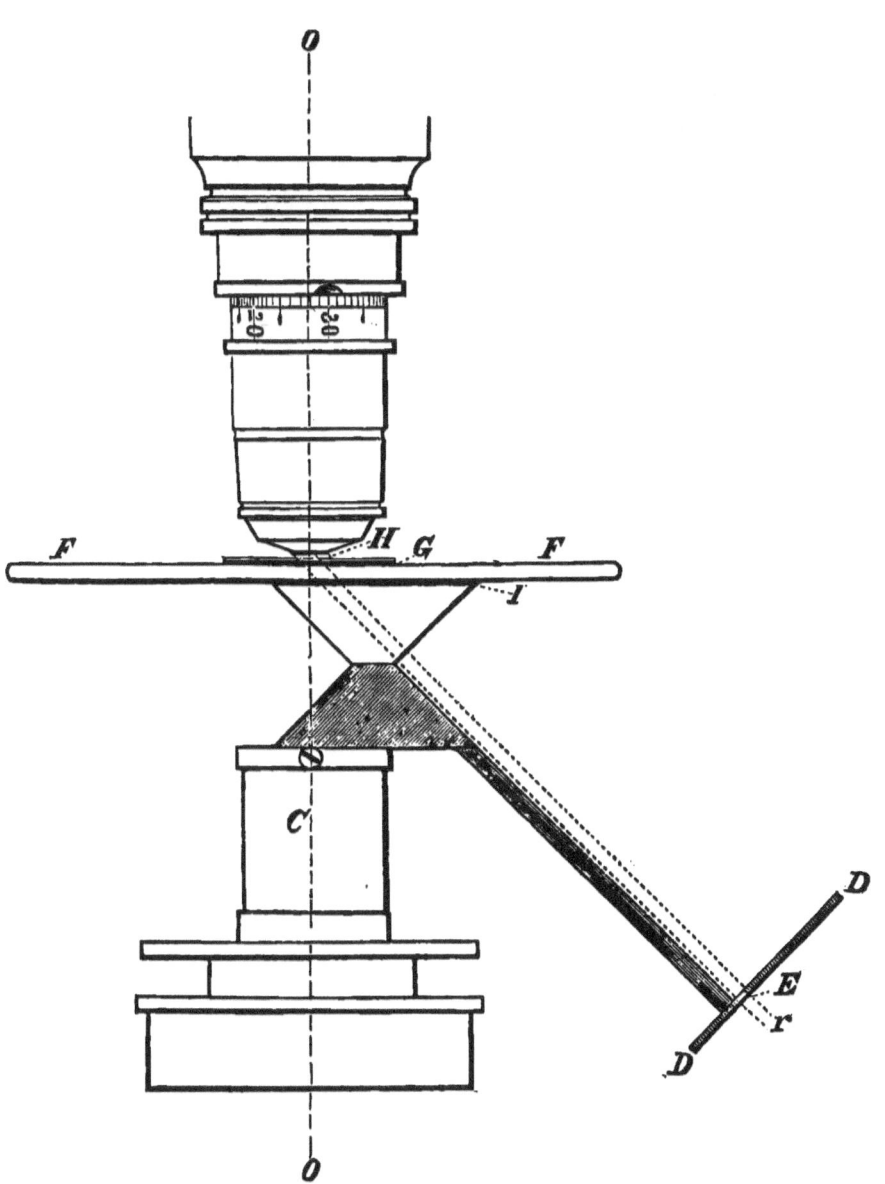

with the slide on which the object is mounted. The details of the plan will be understood by the diagram on next page, in which the glass prism is seen in section just beneath the object slide F F. Just below it is another right-angled prism, of the same dimensions, made of brass; the section of this prism is indicated by dark shading in the diagram. The right-angles of both prisms are truncated, and the facets are cemented together in such a manner that the long sides of the prisms are parallel. The brass prism slips transversely in a groove in the top of a holder, C, which is fitted into the sub-stage of the microscope. D D is a blackened brass screen held in position by two brass arms, one of which is shown in the figure. This screen is parallel to the adjacent face of the glass prism, and has in it a small circular aperture, E, about the size of a large pin hole. The side of the glass prism next the screen is covered with black paper, in which is a corresponding pin hole. The two pin holes are so placed that a beam of parallel white sunlight (r) passing through both will be perpendicular to the sides of the glass prism on which it impinges.

"To use this apparatus it is adjusted in the sub-stage of the microscope, a drop of oil of cloves is placed on the upper face of the prism, the glass slide F F, on which the object is mounted in Canada balsam under the usual thin cover, G, is placed on the stage, and the sub-stage is racked up until the drop of oil of cloves is spread out into a thin layer, I.

"The object being thus arranged, it is evident that if

a beam of parallel solar rays (white sunlight), reflected from a plane mirror, be thrown through the two apertures upon the face of the prism, being perpendicular to that face, it will enter and pass through without refraction until it reaches the upper surface of the thin glass cover G. The parallel rays impinge upon this surface, as is evident from the construction, at an angle of 45° with the optical axis O O. If, now, the medium next above the thin cover, G, be air, this obliquity will be greater than the critical angle, and total reflection of the rays will take place. If, however, the medium next above the thin cover be water, the obliquity will *not* be greater than the critical angle. Refraction having taken place, the rays will enter the water, H; and if an immersion lens of sufficient angle of aperture be focussed upon the objects mounted beneath the cover G, these rays not merely enter the front of the objective, but will form a well-defined image of the object on a brightly illuminated field, which will be visible through the eye-piece of the instrument in the usual way. Of course it is evident from the diagram that with no dry objective, or any immersion objective of less than 90° balsam angle, can anything whatever of balsam-mounted objects* thus be seen.

"Immersion objectives may be divided according to their behavior, with this apparatus, into three classes: 1st. Those with which, since they do not have sufficient angle of aperture to admit the illuminating pencil,

---

* The apparatus can be used, of course, to secure black-ground illumination of suitable dry objects if they are mounted on the slide instead of the cover, as is usual.

nothing can be seen, precisely as in the case of dry objectives. 2d. Those which have sufficient angle of aperture to admit rays of this obliquity, but are incapable of bringing them to an image-forming focus; with these the field appears well illuminated, but the objects are not well defined. 3d. Those which not only admit rays of this obliquity, but form well-defined images with them. To this class belong not merely immersion objectives with the so-called duplex fronts, but others; and I may add, not merely objectives of American make, but some constructed by a well-known English house. As might be expected, the quality of the image formed by the direct rays of the sun thrown through a pin hole at this excessive obliquity varies very greatly in different cases. I will state, however, that I have thus far found at least seven objectives, some of English, others of American make, which define sufficiently well under these circumstances to resolve *Amphipleura pellucida* mounted in Canada balsam. With the objectives which performed best, the field was of exceeding whiteness and brilliancy, but by no means dazzling, the frustule undistorted, and the striæ clean and black on the white ground, very little color aberration being perceived. With other objectives there was more or less color aberration and distortion, both which faults were in one or two cases very conspicuous, although in the part of the frustule most sharply focussed upon the striæ were handsomely brought out. The objectives with which I thus succeeded ranged all the way from one-fourth to one-sixteenth immersion. I will add that the objectives

which resolved *Amphipleura pellucida* under these trying circumstances, when used in the ordinary way with
this or other test-objects, displayed an exquisite perfection of definition which it would be hopeless to expect
to attain with objectives of less angular aperture.

"As it is no part of my purpose in this communication
to provoke ill-tempered discussion of the merits of individual makers, I will not append a list of the results
obtained with the various immersion objectives I have
tried in this way. The apparatus can be constructed
for a few shillings, and those who take the trouble to
use it will soon see to which of the three classes any
particular objective they may test belongs."

Subsequent to the date of the reading of the preceed-
ing communication, Dr. Woodward proposed some slight
changes in the form of his prism. Having had considerable experience with the prism as now used, we append
the following description, and also the manner of working the instrument.

Essentially, it consists of a triangular prism of crown
glass. In the form adopted by Dr. Woodward the obtuse angle is 98° and the two acute angles 41° each.

The prism may be used unmounted, by simply attaching the same to the under surface of the slide containing
the objects to be examined, a drop of glycerine or oil
of cloves serving to secure optical contact, and at the
same time acting as cement to retain the prism in place.
Notwithstanding this is the arrangement employed by
many observers, it is but a faulty plan, in fact, a regular

"make-shift" arrangement. With the prism thus mounted any movement of the object-slide will of course cause a corresponding movement and decentering of the prism; furthermore, such object-slides can only be well examined when posed in nearly the horizontal position on the stage; it often, too, occurs that the intermediate drop looses its hold, suffering the prism to slide or even to loose its attachment entirely; and then, again, I have frequently got the very best work when the facet of the prism was slightly depressed from the under surface of the slide.

It is far better, then, and for the reasons given, to have the prism mounted, and to those who propose to adopt my method of illumination, I will say that much depends on the proper mounting, and that any piece of sub-stage apparatus which shall impede the passage of rays from the lamp to the mirror, or from the mirror to the prism, will defeat the maximum working of the latter. Hence, as a rule, it cannot be well mounted in the usual sub-stage, the latter obstructing too much light.

After much experiment, and with the hearty co-operation of Mr. Sidle, I am now in possession of the Woodward prism, suitably mounted and adapted to the "histological" of Mr. Zentmayer, as will be seen from inspection of the cut on the following page.

This accessory, as above delineated in plan and section, is easily placed in position on any of the histological stands. Provision is also made for centering in a lateral direction. The prism can also be revolved so as to use either angle. It can at will be raised or lowered, or clamped in any desired position.

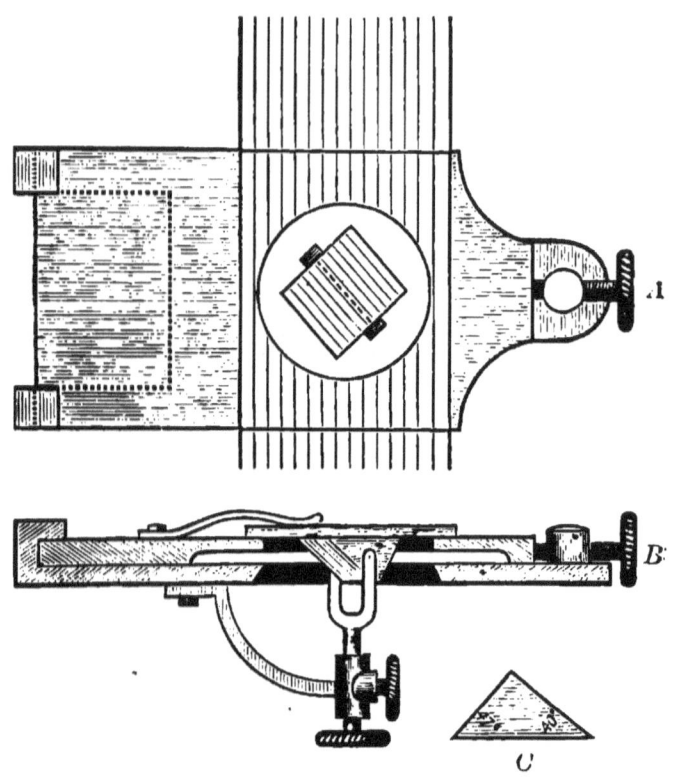

a. Vertical view.
b. Sectional view.
c. Prism three-fourths full size.

In the adaptation of this useful accessory to the acme stand Mr. Sidle has been singularly fortunate.

This neat and compact little device screws, as I have had occasion, on a previous page, to state, into the well-hole of the acme stage. It is thus "on and off"

in a moment's time.   All the necessary motions are provided for, and it may be depended on for first-class
performance.

From what has already been stated, it is almost needless to repeat that when either of the mountings described are to be used on the histological or acme stands
the sub-stage must be removed.   It is quite possible that
the mountings presented may, with slight modifications,
be fitted to other microscope stands.

· The form of prism (angles 98°, 41° and 41°), as suggested by Col. Woodward, will be found to do very
acceptable work.   It occurred to me, after having had
considerable experience with this accessory, that there
was nothing gained by having the acute angles equal,
but on the other hand advantages would be insured by
an inequality of these angles, either of which might be
used as occasion required.   I therefore begged Mr. Sidle
to make me a prism with an obtuse angle of 93°, one of
the acute angles to be 47°, the other to be 40°, and this
is the form I have adopted, believing it to be the most
seviceable arrangement yet proposed, and especially
adapted to the general run of modern wide-apertured
objectives.   To provide one's self with two or more
forms of prisms will not involve serious expense.   Either
of the mountings I have described can be modified so
that the prism can be removed and others substituted
in its place.

To use the Woodward prism—we will suppose on the
acme stand—first, screw the accessory into the well-
hole of the stage securely; next, place the slide to be

examined on the stage. With a one and one-half or two-inch objective (having raised the prism so as to contact nearly the under surface of the slide), focus entirely through the body of the prism, so that its lower edge may be seen in the field, and by turning the prism in its mounting make this line, by estimation, as nearly vertical as possible. If, perchance, this line appears to the eye considerably to the right or left of the centre of the field there is no harm done; but you must make a note of the fact; this would indicate, however, that either the stage is out of centre, or the mounting of the prism has been injured.

Next, rack back the objective and remove the slide, place a drop of glycerine on the top face of the prism; replace the slide and raise the prism so as to make contact with its under surface. Having made this contact exactly, depress the prism, say, about one-fiftieth of an inch. Focus again with the low power, and bring the lower edge truly vertical, as before instructed.

Remove the low power and substitute the wide-apertured objective, and by way of illumination provide a small kerosene hand lamp, the flame of which ought not to be higher than two-thirds the distance from the table to the stage of the microscope; remove, also, the substage.

Now, if the lower edge of the prism was seen to the right of the field place the lamp to the right of the stage; on the contrary, if the edge was seen to the left, place the lamp to the left; and in either case swing the mirror *away* from the lamp, placing it so that the graduating wheel shall read at about 34, 35 or 36 degrees.

The lamp being passed to the right or left of the stage, as the case may be, and about four inches distant therefrom, bring the edge of the flame to the mirror; now move the lamp to or from the front edge of the *table*, so that the edge of the lamp frame, prism and centre of mirror shall form approximately a straight line parallel with the front edge of the table,

From the position described it will be seen that the ordinary sub-stage would be entirely in the way; hence the necessity for its removal.

Making immersion contact, focus your objective, and, without changing the position of the swing bar, manipulate the mirror so that the field may be nicely illumiated, select your object, which we take it for granted will be some difficult lined test.

Next, interpose the large bull's-eye condenser (flat side to the lamp), thus concentrating light on the mirror; adjust the object glass.

It will be well now to try the effect of various degrees of obliquity, remembering that any considerable movement of the swing-bar will necessitate a new adjustment of the condenser. A slight change, too, in the position of the lamp will sometimes be attended with excellent results, keeping, in all cases, however, its edge to the mirror.

The above includes the author's method of working the Woodward prism; but as this accessory bids fair to come into general use he will now traverse the ground over again, feeling sure that many there are who will not object to some further discussion of the subject.

In the initial attempt to use the prism, the observer should select an object (balsam mounted) with which he is tolerably familiar. The collar adjustment, also, should have been previously ascertained. The next step is to decide on the proper position of the radial bar, i. e., its distance from axis, the extent of this distance will be demonstrated by the illumination becoming too feeble, the images, also, becoming generally unsatisfactory. The . remedy is, in such cases, to cause the radial bar to approach nearer to an axial position, until the field can be sufficiently lighted and the object displayed with tolerable vigor. Thus the operator has the means of " gauging " his objective. The mirror being properly posed, it remains to obtain the best possible illumination, which is effected by slight changes of the mirror, condenser, and finally the lamp. When things are generally about right, a little movement of the lamp (grasping it firmly by the bowl), sometimes twisting it to the right or left, so as to get the flame exactly edgewise to the mirror (which is best determined in this way), will result in very nice effects, bringing out the striæ on such tests as the Moller test-plate in a very satisfactory style.

Resolutions of difficult test objects are accomplished with this illuminator in a very handsome manner. It is an easy instrument to use, and will adapt itself kindly to the objective, of course; the higher the balsam angle of the object glass, the better the definition. It has, in its ease of adaptation a decided advantage over the Wenham forms, and in the comparative examination of objectives, the wedge illuminator is an exceedingly handy accessory. In the determination of the collar adjustment corresponding to the point of maximum aperture, the same holds true.

We have spent two or three entire evenings in the attempt to determine which of the two illuminators is the most effective; and our experience leads to the conclusion that the "reflex" is somewhat the superior instrument in the resolution of the most difficult lined tests; nevertheless, we are glad to give the newcomer a place in the accessory box, and expect to make it very useful. It is easily made and mounted, and ought not to be expensive.

Another matter closely allied to the new illuminator may as well be mentioned here. Learning that the Messrs. Spencers had just completed a new one-fourth inch objective, which was to be sent to the Paris Exposition, we wrote to these gentlemen asking the loan of the glass for examination, the Messrs. Spencers responded promptly, and it occurred that we received the one-fourth and the new illuminator the same day. We were thus enabled to put the illuminator to practical use at once, in this manner: First, we took our one-sixth, working it with the illuminator over the No. 20 of the Moller plate getting the radial bar as far from axis as the objective would allow and preserve a good display of the striæ. This done, we substituted the one-fourth in place of the one-sixth, keeping the illumination, etc., carefully in the same position, (the cover of the plate was as well adapted to the one glass as the

other.) We found at once, that in order to obtain sufficient light, and retain the general vigor of the image,
it was necessary to approach the radial bar to the axis
and the required movement of the latter was quite perceptible. It was therefore accepted that the one-sixth
had the higher balsam angle. The question then turned,
as a matter of course, on the respective working distances; that of the one-sixth was known. It remained,
therefore, simply to measure that of the one-fourth,
resulting as follows: The working distance of the one-
sixth is twenty-four–thousandths of an inch; while that
of the one-fourth was found to be thirty-two–thousandths of an inch, a difference of 33 per cent in favor
of the one-fourth. Thus it will be seen that in this
instance the question as to superiority may be further
taken under advisement. We relate this bit of "practice" with the illuminator in illustration of the preceding remarks.

One thing was proven even by the above experiment,
to wit: Having our tenth on hand, held in reserve for
especial demands, we would greatly prefer the one-
fourth as an intermediate glass. This fact is too obvious
to need further comment, and, in general, we are glad to
add that the new one-fourth of the Messrs. Spencer is
indeed a lovely glass, and if properly exhibited in Paris,
will be an honor to the talented makers.

The modified illuminator above described (obtuse
angle, 98°, acute angles, 41° each) will work very well
with objectives having wide *air* apertures only; hence,
like the modified "reflex," it will work over dry mounts,

in such cases, in common with the modified "reflex," its action, to some extent, is crippled; nevertheless, nice resolutions are to be obtained with either instrument, with either of which we are able to instantly display the transverse striæ of *amphipleura pellucida*, *frustulia saxonica*, etc. The simplicity of Dr. Woodward's device, its ease of working, and the facility it affords for the comparison of objectives are in themselves strong points in its favor, and to these may be added the satisfactory character of the resolutions obtained. We gladly accord to Dr. Woodward our appreciation of the value of his illuminator.

We are informed by General Cox that the pin hole apertures are only used when working with sunlight illumination; they also serve a useful purpose as an assistance in measuring with greater precision the obliquity of the illuminating pencils employed, thus enhancing the value and capacity of the instrument.

TOLLES' TRAVERSE LENS.

It now remains to present to the reader the Traverse Lens, devised by Mr. R. B. Tolles. The following is the inventor's own description of this valuable accessory, and is taken from the *American Journal of Microscopy*:

"With the advent of objectives of increased interior angle aperture, the indispensableness of equivalent accessory means for the illumination of the object became immediately evident.*

---

* See "M. M. J.," July, 1871, p. 38.

"In my first construction of such object-glasses I therefore required to provide means which proved so suitable that I have adhered to their use to the present time.

"The first appliance was a deep plano-convex lens, centrally mounted below the object, and having its centre of curvature in the object place. Afterwards I adopted a plano-cylindrically convex lens, equal to a hemisphere less the thicknsss of the object-slide, which was placed in immersion contact with the base of the slide, so that the object itself formed the centre of curvature of this illuminating lens. Around the convex surface of this central lens moved a shutter to regulate and limit the access of light, and it was provided also with a small plano-concave lens which, applied by its concave to the convex surface of the larger lens by immersion contact, cancelled the refracting surfaces and allowed a perpendicular *beam* of light to reach the suitably immersed object without refraction.*

" The device in a more complete form is represented in the annexed figure, where P is the basilar plate of the whole traverse system, having a circular groove and track in which the carriage, C, moves. On a projecting arm, A, of the carriage, C, are mounted whatever appliances are to be used to modify or direct the light upon the traverse lens, T, in the direction of the object at the centre of the system.

" In the figure, the concave lens, N, is shown in position on the arm. Thus situated, the interior convex

* M. M. J., May, 1873, p. 213.

and concave surfaces being of no effect, the two exterior
plane surfaces of *the traverse system* constitute it a prism,
and every slightest movement of this concave facet lens
on the traverse lens, T, would *give a different prism* to
infinite variety. In this arrangement the concave mirror can be used in the ordinary manner and condense
light enough upon the object for all ordinary purposes.
The full interior aperture of a dry objective would be
reached at the véry convenient obliquity of 41°; *i. e.*,
at less than the critical angle, or angle of total internal
reflection between crown-glass and air. L is a double
convex condensing lens, that may be placed at about its
principal focal distance from the object.

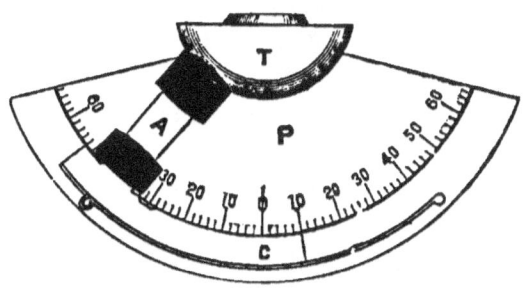

"For a condenser, with the size of apparatus as drawn
in the figure, a simple lens of 1¼ inch focus, and about
ten (10°) degrees of aperture, is convenient, and if the
lens is movable along the arm, A, it can be focussed
readily on the object, the position being fixed by inspection. This would be well for parallel rays. If diverg-
ing rays are used another lens of two or three inches
focus, mounted on the arm, A, will conveniently take

up the rays from the radiant at the distance of the focus of this supplementary lens.

"The plate, P, is graduated on its circular edge, as in the figure, to two degrees, and the arm, A, has a swing of seventy degrees of arc each way from the axis of the microscope. An index-line is marked on the bevelled edge of the carriage 10° from the axis of the condenser, which must be added to or subtracted from the real obliquity of the illuminating rays.

"It is obvious that any observation made and duly recorded as to its conditions, as of obliquity of incidence of illuminating pencil or ray, form of the pencil or beam, focal length and distance of the condenser, such observation could be successfully repeated. The record of the obliquity of the most oblique rays reaching the object directly, and giving view of it at the eye-piece with luminous field, would express the 'balsam' aperture, or more correctly, the half interior aperture of the objective when the front lens of the objective and the traverse system are of glass of similar refraction.

"Having thus the 'balsam' angle, we readily calculate or learn the corresponding angle for glycerine, or water, or any medium of which we have the index of refraction. A corresponding notation, perhaps for air, might be engraved in juxtaposition on the basilar plate."

CHAPTER V.

Ordinary daylight is the cheapest; and for a great many purposes the microscopist will find it amply sufficient. It will be found a great convenience to have the light enter the room considerably to the left of the microscope, in which case we naturally adjust the mirror with the right hand. Placing the instrument directly before the window is objectionable, and such a position should be avoided if possible. The quality, as well as the quantity of daylight illumination will, as a matter of course, vary with the particular aspect of the day. In bright sunny weather the light from a white cloud, as has often been recommended, is pure and pleasant to work with, and can be used with tolerably high amplifications with good success. In cloudy, rainy weather it is still quite possible to work with powers up to, say 200 diameters. The recent introduction of the swinging sub-stage has worked somewhat of a revolution in our own practice. For years we have steadily eschewed the achromatic condenser as being a costly and inconvenient affair, making more " bother " than it was worth. The principal objection I had to urge against its use was that it was a *fixture* beneath the stage, thus preventing me from varying the obliquity of the illumination at will, as I desired, and, as a rule, practically I got better effects without than with it. The former condensers were generally of short focal length, and of considerable aperture. In the late stands having swing substages, it being possible to swing sub-stage and condenser together bodily, there seems to be no further use for condensers of wide angles, while on the other hand one is enabled to use in the place thereof cheaper and much less expensive instruments, and the lower the angle the better, and one need not be very particular as to the matter of achromatism.

On commencing the use of the little Histological, it occurred to the author (and probably to scores of others) that its swinging stage was a strong invitation to experiment again with sub-stage condensers, not for the purpose of resolving difficult tests by extremely oblique illumination, for in this work the achromatic condenser is of no manner of account, but, *per contra*, it seemed obvious that by the use of a narrow apertured lens placed below the stage, and so arranged that its inclination might be changed at will, good effects might be secured in two directions: First, by the concentration of a narrow cone of light immediately upon the particular portion of the object under examination, thus enabling the observer to sharply illuminate a certain point of his object, and with less danger of drowning out details in a general flood of light. Secondly, such a contrivance would do good service by daylight in dark and rainy weather. It required but few experiments to demonstrate that there was force in the above reasoning, and the next thing in course was to ascertain what par-

ticular form of condenser would be the best adapted for
the purpose.

In determining how far it would be practicable to cut
down the angle of the condenser, thus reducing the
illuminating cone of light, we have made countless experiments, while the low stage of the Histological rendered it imperative that the focal length of the lens
should be such as would best accommodate the little
stand. To sum up all these trials, we find that the
cheapest inch objective made by Mr. Gundlach, or the
inch of the Messrs. Beck's "National Series," are, either
of them, well adapted for the purpose. Mr. Gundlach's
inch has a rubber front which can be removed, while
the setting of the Beck "National" is extremely short,
and thus suited in this respect for the purpose.

This, then, is the author's arrangement for work with
low or moderate powers by daylight illumination, and
the condenser described has become almost a fixture.
In the darkest days there will be plenty of light, using
the concave mirror, while in bright, sunny days the
plane can be substituted. The general amount of illumination can be changed at will by merely raising or
lowering the sub-stage, and the nicest effects in the way
of definition obtained. The swing-bar can also be
placed so as to afford central illumination, or it may
(condenser and all) be swung laterally up, say to an
angle of 40 or 50 degrees from the axis; and it further
remains to say that either of these cheap objectives are
real good, honest glasses for the money.

SUNLIGHT.

In the study of very minute and delicate structures requiring the utmost separating or resolving power of the objective, remarkable effects are to be secured by condensing sunlight *on top* of the object by means of the concave mirror, the object being mounted with a cover in the usual way. The objective used should of course have wide aperture. The mirror being posed slightly above the level of the stage, the sunlight is thrown on the surface of the cover, and making a very acute angle therewith. Although not absolutely necessary for this purpose, those stands furnished with swinging sub-stages, allowing the mirror to rise above the level of the stage, are extremely handy and convenient. By the employment of this illumination in conjunction with object-glasses of wide angles, the most difficult diatoms, such as *amphipleura pellucida, frustulia saxonica,* etc., are easily and forcibly displayed. The advantages attending the use of monochromatic sunlight, as obtained by the intervention of the cupro-ammonia cell, or a plate of blue glass, have long been known. This illumination is procured most easily as follows: Cut with a diamond, or the point of a file, a small piece of the blue glass roughly to fit the *cap* of the eye-piece, so that when the cap is restored to its place the blue glass shall be between the eye and the eye-lens of the eye-piece, and the light is thus modified before it reaches the eye. This is the handiest method of obtaining monochromatic illumination we have ever tried, and the resolutions are quite as strong and effective as when the cupro-ammonia cell is used in the usual manner. In working with sunlight by either of the methods described, care should be taken to exclude the full strength of the solar beam; that is, if the sun be clear and bright. Too much light, supposing the manipulations are tolerably well attended to, will be manifest by the appearance of a multitude of diffraction lines, and these as a rule may be recognized by their extending beyond the object observed. Under very high amplifications, involving the use of powerful eye-pieces, we can of course help ourselves to a little more of the solar beam. When the sun is very clear, the beam being condensed on the top of the cover, as above described, there is danger sometimes, if the object be balsam mounted, of the heat starting the balsam. In this way we once ruined a Moller probbe plate. A very little attention will, however, provide against accidents of this nature.

ARTIFICIAL LIGHT.

For the ordinary purposes of the microscopist the St. Germain or German Student's Lamp, C. A. Kleemann's patent, or a similar lamp made by the Cleveland Company, will be found quite satisfactory. This style of lamp is too well known to require any extended description. The flame is bright, clear and intense, and its height can be changed at will. It is easily kept in order, and has the advantage, too, of being well adapted for ordinary household purposes. The breakage of

chimneys has been a serious objection to its use; a brand
of chimney known in Cleveland as the " Crown " (each
chimney having a crown ground in the glass) seem to
be very free from breakage. Non-combustible wicks
are to be obtained, fitting the *Kleeman* lamp. These
are clean and handy, obviating the necessity of occasional cutting and trimming; but to our mind the light
is not so intense, and therefore we prefer to use the old
style of wick. These lamps burn very steadily, and are
not easily affected by occasional drafts, and this is a strong
recommendation in their favor, as is also the case with
which they are kept in order. For investigations of
exceedingly difficult objects the circular wick is not so
well adapted, and recourse must be had to lamps carrying flat wicks. The best lamp we know of, of the latter style is the Mechanical Lamp, manufactured in New
York City. The lamp stands about ten inches in height.
The height cannot be changed, and this is an objection.
It burns kerosene oil, without any chimney. The body
contains a movement which, on being wound like a clock,
drives a blast wheel, and thus supplies a current of air
at the point of combustion. Although there is a peri-
odicity noticeable in the burning of this lamp, nevertheless the flame is very steady, is very intense, and
superior to gas. Like the St. Germain, this lamp is
very handy to have in the house, and it takes but little
trouble to keep it in order. The movement should be
cleaned once a year, and any one possessed of fair mechanical skill will be competent to do this. While
burning, the clock-work makes scarcely any noise. An-

other form of this lamp has the movement placed in the case flatwise, thus allowing the flame to burn within three or four inches of the table. The lamp is thus rendered very handy for use when direct light is wanted.

The author has found, as a result of thousands of experiments, that the very best artificial light for the purpose of the microscopist is only to be had from a small but very intense flame. The smaller the flame the better, owing to the fact that there is less light diffused. We therefore use and strongly recommend the smallest kerosene hand-lamp procurable, and fitted with a well-behaved burner of the smallest capacity. If possible let the lamp bowl be so low that the flame will be, say three' or four inches only above the table, thus adapting the lamp for use by direct light. On other occasions the lamp can be supported in a more elevated position. With a little lamp of this description, in proper order, all the most difficult tests known to the microscopist can be well displayed, provided, obviously, that the objective, etc., shall be competent for the work. It is of importance, when any lamp provided with a chimney is to be used, that the latter be kept scrupulously clean, especially from a whitish film that forms on the interior. A chimney may appear to be perfectly clean while cold, but when heated the aforesaid film can be detected, and should be removed, if delicate observations are in hand, in which case it will be well, too, if the wick be three or four weeks old, to remove the same and substitute a fresh one. Even in the case of the small pattern of lamp recommended there will be no occasion to force the combustion to the fullest extent such a burner will afford. A flame with the lamp burning at one-half its capacity will be amply sufficient, and even this would be too much for the proper display of some of the most difficult tests.

Attempts have been made to modify artificial illumination by the introduction of blue tinted chimneys, white ground illuminators, etc. We have patiently tried the entire list, and reject them all, from the fact that there is no real advantage secured by their adoption which cannot be obtained in a simpler way without them. The neutral tint " Light Moderator," so called, is a pleasant thing enough for use with moderate amplifications; yet there is nothing seen with it that cannot be as well shown without it.

The blue tinted chimney cuts down seriously the intensity of the lamp illumination to an extent which will defeat the resolution of any severe test, while, on the contrary, any and all work with the lower powers can be as well accomplished without its aid.

The reader has thus before him all the various kinds of illumination we use. A great deal of the professional routine of work not requiring, as a rule, the employment of the highest amplifications (such as the examination of urinary deposits, malignant growths, etc.,) we try as far as possible to accomplish in the day time, and by the use of diffused daylight. If the sun happen to shine, and it be desirable to cross-question some preliminary examination under the highest powers, we generally use the sunlight condensed on the top of the cover, or perhaps with the aid of the bit of blue glass in eye-piece. For work at night we employ at times all the lamps we have described. Should the routine examinations be prolonged into the evening, we use the German student's lamp for preliminary work, the same as we use diffused daylight in the day time. But should the higher amplifications become necessary, we bring the mechanical or the little hand-lamp into play. The German student's will still do service in the lighting up generally of the work-table at intervals. For the showing of such objects as the Nos. 18, 19, and 20 of the Moller plate by lamplight, of course the little hand-lamp, or the lower model mechanical is imperatively employed, especially when the Wenham " reflex " illuminator is selected.

There remains yet another method of sunlight illumination which will be found useful at times. I refer to the use of the " reflex " illuminator with direct sun-light. In this case the solar beam can be received through a closed window (quite a boon in the winter season) and reflected from the plane mirror. This illumination is only suitable for work with wide apertures, and ever the most minute objects, and the mount must be free from surrounding objects of a coarse character, else, from the extremely oblique character of the illumination these stronger and coarser objects will project their strong shadows across the field, causing nothing but confusion and chaos. With the genuine form of the Wenham " reflex " an epithelial scale would hardly be recognized were there several in the field. The princi-

pal advantage in the use of the "reflex" with sunlight
is in arriving at a knowledge of surface markings, and
for this purpose it is indeed very valuable. Thus working the "reflex" by sunlight, the mirror must be manipulated so as to produce the same effects as have been
described by moving the hand-lamp and conversely.
The mirror may be substituted for the hand-lamp when
working in the evening, but the most favorable results
are obtained with the light direct. This reflex and sunlight illumination is especially desirable when one wishes
to trace out structure situated in one particular plane,
to the exclusion of that lying in adjacent planes. In
the general squabble to produce the so-called penetration, this very important item has been lost sight of.

We are now ready to consider a matter which has
been alluded to on a preceding page. It has been
already stated that the maximum performance of *adjustable* objectives can only be secured when such object-
glasses are worked at the point of their maximum aperture, and that this point is by no means a fixture but
varies with different objectives. Every observer should
then ascertain for himself as to the proper handling of
his object-glasses in this particular. Methods will now
be given which, although but approximate, are sufficiently precise for the use of the practical manipulator.

For the purpose of testing the point of maximum
aperture for object-glasses having apertures, say from
40° to 175°, proceed thus: Place the objective in position on any good stand having a thin stage and mirror
attached to radial arm. Commence by focussing any

suitable object on the stage, with the mirror in a central position, the collar of the objective being set at the extremity of its range. Now, by degrees, swing the radial bar carrying the mirror, meanwhile adjusting the mirror so as to secure all the illumination possible, just as would occur in arranging for greater obliquity of illumination, until the obliquity of the light becomes as great as the objective will bear; *i. e.*, until the greatest degree of obliquity has been obtained that will secure a tolerably well-lighted field. Now move the radial bar a little, and but a little further from axis, meeting this change by the proper manipulation of the mirror, and so as still to secure all possible light. The object now ought not to have more than one-fourth the usual illumination, but should nevertheless be distinctly seen. Next, revolve the collar and notice the effect. If you get less and less light as the collar is turned towards the other extremity of its range, it would show that it was already at the point of its maximum aperture; on the contrary, should you get more light, it will be apparent that the aperture increases as the collar is turned, and thus turning the collar by degrees, move also the radial bar still further from axis, manipulating the mirror as before, and to the same end, and so proceed as long as the change of the collar gives more light. You have then, by simple inspection of the position of the adjustment, a tolerable idea where the maximum of the objective is to be found. Note this: Now remove the object, place the stand in a horizontal position, and, without changing the adjustment of the objective, proceed to measure its angle by the method previously given. Note again the angle obtained. Next, change the collar adjustment a division or two, and again measure the aperture. Comparing results, it will become obvious which of these two positions of the collar corresponded to the larger angle. Should the latter measure prove the least, it will be necessary to reverse the movement of the collar, placing it a division or two from the previous position but in the reverse direction, and by a few measures of this kind, which, by the way, are quickly accomplished, the point in the collar adjustment corresponding to the maximum aperture on the glass will be ascertained with considerable precision. In the method just described the primary object was to get an approximate idea as to the point of largest aperture, and with the least outlay of time, and subsequently, by actual trial, to arrive at a more precise determination. The whole process involves but little outlay of time, ten minutes being quite sufficient for the purpose.

With objectives of high balsam angles it will be necessary to employ the genuine Wenham " reflex " illuminator (angle of facet 26°). With this instrument proceed as has already been advised until, by the lateral movement of the lamp at either the extremes, right or left, the illumination commences to die away, the field being blue or red, according to the position of the lamp. It will generally be a saving of time to start with the collar of the objective at " closed." Having found the best position for the lamp, as we have before directed, move it still a little further laterally until the field of the instrument shall only be illuminated sufficiently to enable you to see your object distinctly. Now, keeping things thus, revolve the collar, and notice the effect on the illumination, and thus, as in the case already presented, you have the means of judging as to the aperture of the objective. And as an example I now relate a bit of experience not twenty-four hours old: We have just had in hand an objective claiming to have high balsam angle, and we desire to know something about it. First, we look to its working distance and find that it will work through covers one-fiftieth of an inch thick, its distance is therefore ample. Applying the Wenham "reflex" we test as to aperture, and precisely as has been above described, thereby learning that its greatest angle occurs when the systems are at "closed." We find, too, that as the collar is revolved towards the "open point," the angle goes down rapidly. We therefore conclude that to work this glass at its maximum performance it will be necessary to use covers thick enough to cause the objective to "correct" at or near "closed." It will take but a moment to try the actual experiment, and to see if theory holds good in practice. For this purpose we place the "reflex" in position and the No. 20 of the balsam Moller plate on the stage, making immersion contact with water. Next, we attempt the resolution of the shell, and with the best manipulations at our command succeed in getting but a tolerable show of the striæ. I have the blue field in hand, with the lamp at the extremest point to the left: the best display being thus obtained, as the lamp was thus shoved to the left the definition was improved, but we were compelled to desist from this movement owing to the loss of light, and were therefore content with the lamp as far to the left as was possible without sacrifice of the illumination. Of course the objective was adjusted with all possible care, the collar standing within *three* divisions from " open point."

We now carefully raise the objective, and removing the water with a bit of blotting paper, we substitute a drop of glycerine, focussing and adjusting the glass again with the glycerine intermedium ; the glass now adjusts at nearly *closed*, the collar having made nearly *two* full revolutions from its former position. It is further obvious that we have now more light generally; we can, too, move the lamp to a greater distance right or left without loss of illumination. In fact, things in the tube have a sunshine appearance that is very acceptable. We now attempt again the resolution of the same shell, using the blue field as before. Finding that the lamp will bear to be shoved further to the left than before. And now, even before arriving at the limit of light, *i. e.*, the lamp not so far to the left as we might place it, we are rewarded by a splendid display of the transverse striæ, this, too, with illumination, I was going to say in excess, at all events enough to allow the use of the one-half and the one-fourth solid eye-pieces.

We have thus described this experience taken from our private practice, giving the actual results obtained. Well, now, suppose that with the Wenham reflex the experiment had turned out a total failure; *i. e.*, that we could get no light through the objective, showing that in this case the accessory was a *reflex* and no mistake; or, suppose that our best efforts were only rewarded by a dim view of the diatom seen doubled up endwise amid a plexus of indeterminate undefinable diffraction lines, the entire field miserably illuminated, of a muggy, smoky, dingy yellow, as if a piece of yellow flannel had been used in place of the lamp, and that we could only get this much in just one solitary position of the lamp; then the experiment is quite as interesting, quite as valuable, and *if made in due time will amply repay its cost.*

As a matter of course, the lower the balsam aperture the lower will be the grade of its work with the " reflex;" nevertheless, the point of maximum performance can be ascertained by the method above given, and in the testing of object-glasses I make it a point to look after the balsam angle and the point of maximum aperture at the same sitting, the only additional trouble involved being the change in the thickness of cover, or, as in the instance named, substituting glycerine in place of water.

We have entered into these details, feeling assured that the facts are worth knowing, and that many there be who have not given these things due attention. Of those who have visited us, eight out of ten saw the working of the " reflex " for the first time, while without exception all have seemed greatly pleased and interested with such comparative experiments as we have just described. Let the reader rest a moment here and I will relate a little incident:

Not long since I was honored with a call from a gentleman who had put himself to some little trouble in visiting Cleveland. Said he: "I have been for some time desirous of visiting you; I have read all your articles in the journals, and have been permitted to read your letters to our mutual friend, —— ——. I want to know something more about the "balsam angle" business; ditto, about the "180°"; I confess these extreme apertures seem to me "impossible"; and then, again, we have high authority that the true aperture of any object-glass cannot exceed 120° and the assertion is fortified with reasoning that I cannot well dodge. I have brought with me an excellent glass purporting to have corrected angle up to 140°, and would like to have some comparisons made with your own. I am after the facts, and have no personal bias in any direction," etc., etc. He also went on to state that Mr. Wenham had expressly asserted that direct light could be obtained with the "reflex." At his suggestion I showed him the No. 18 of the balsam Moller probbe plate illuminated with the genuine "reflex." The field was brilliantly lighted with just enough of the blue in to take off the intensity of the glare. The shell appeared without sensible distortion, edges sharp and clean, and with a full stand of lines from end to end. Revolving the stage so as to place the *saxonica* in a diagonal position, we had little difficulty in obtaining simultaneous views of both transverse and longitudinal striæ, thus cutting the valve into checks or squares. The same little hand-lamp was used, and we had nice shows with eye-pieces up to the one-fourth inch.

My visitor was delighted, nor did he attempt to conceal his delight. "Now," said he, "just keep all things just as they are, but take off your glass and put on mine." The same was accordingly done, and the result was that we could not see the diatom at all, nor could we, by the best possible manipulation of the lamp, see it well enough to recognize it. I suggested the possibility that the glass might not be truly centered, and thus to some extent be defeated. Attention was therefore given to this, but without avail. My friend's objective positively refused to have anything to do with the reflex illuminator.

Now, I have found, by countless experiments, that of two objectives, the one working well with the genuine "reflex," and another refusing to so work at all, the former will be far the superior glass for any and all work; a fact which, after a little subsequent experiment my visitor was not slow to accept. I am of course referring to objectives generally known under the appellation of "high powers."

Thus it will be seen that the Wenham "reflex" is for any of these purposes quite a handy and effective little instrument, and ought to have its place in the accessory box of every microscopist. It will serve, too, in its legitimate capacity as designed by its inventor, i. e,, as a "dark ground illuminator;" but herein will be found its least value. To return directly to our subject: I cannot too strongly recommend that every one interested in microscopical work requiring the employment of high amplifications with fine defining power, should study their objectives, and if for the sake of practice only, those of their friends to which they can have access. There is no time lost in this occupation; on the other hand, it will usually result in economy of time. For instance, if your glass totally refuses to associate with the genuine " reflex," you are hereby informed that such a glass is totally unfit for any such purpose as resolving the last numbers of the Moller probbe plate, or for any kind of duty requiring the recognition of lines as close as 80,000 to the inch. Thus, by the method described you can, in less than ten or fifteen minutes time, settle definitely any such question as to the capacity of your object-glass.

This matter is suggestive, and with the reader's permission I desire to " switch off on a side track " again, for we have " an ax to grind." We often, yea, almost every day, hear those who regard the study of objectives as one worth attention, roundly condemned by workers in natural history, biology, etc. The former are said to be " only diatom crackers," who do nothing but fool away their time over difficult diatoms, and are said to have angular aperture " on the brain." Now, reader, when you shall be permitted to peep behind the curtain as often as I have had the chance in the past (which, by the way, I have *improved*), you will find that the very gentlemen who make all this hue and cry are the very ones who have been and still are " fooling away " their time. You will find, as a rule, that each and every one of them have their little cabinet of " difficult tests," over which they spend (sub rosa) night after night in an absurd attempt to display the hateful markings; an attempt, too, as futile and puny as that of the child who cries for a piece of the moon. And why? Simply because the object-glass employed is not suited to the work in hand. Nor is the picture overdrawn, as more than one lady can attest, or the "looker on in Venice" vouch for.

To all such, to all who value the microscope as an aid to scientific investigation, let me urge the importance of studying well the nature, capacity and capabilities of the objective, and to this end, and in the special line we have been discussing, the Wenham "reflex" will prove itself a valuable and important accessory, and a time saver of the very first water.

CHAPTER VI.

Practically, for the past four years, we have confined ourselves to the use of four object-glasses, namely, an inch or two-thirds inch of 45° or 50°. A one-half inch of 38°. A one-sixth immersion, balsam angle ranging from, say 87° to 95°, according to the position of its collar, and a one-tenth immersion having a constant angle of 100°. Of the last two glasses, the one-sixth has a working distance of one-fiftieth of an inch. The one-tenth will work readily through covers one–one-hundredth of an inch thick.

The orthodox theory has been, and I suppose still is, that each worker ought to select his stand, objectives, accessories, etc., with special reference to the particular line of investigations he may elect to pursue; and since, as before intimated, there may be more or less force attached to such a platform, I neither accept it nor reject it; nor am I "on the fence," halting between two conflicting opinions. Without going into any special discussion of the *pros* or *cons*, we will proceed to state the character of the work in which we have been engaged, accompanied with a recital of the special methods, etc., employed.

First, we use the microscope constantly from January to December in examination of urinary deposits, and for the study and detection of malignant growths. In conjunction therewith the aid of medical chemistry is constantly sought, especially in the diagnosis of renal diseases. Work of this description is continually on the tapis "all the year round." In addition to this we use the instrument as a necessary aid in our daily lectures, and for the private instruction of students at the college on matters pertaining to our Chair of Histology and Microscopy. Besides these duties we have more or less private instruction entirely outside of the college to attend to. The range of work then to be accomplished is by no means a narrow one, and anything in the way of instrumentation that would assist either teacher or pupil will find at all times in my laboratory a ready market.

We thus state the character of the work we have in hand, and also the instrumentation employed for its accomplishment; perhaps a leaf or two from our daily practice may prove acceptable, and with this hope we proceed:

We have said that a large amount of our work is over urinary deposits. For this purpose we use the "histological" stand of Mr. Zentmayer, the tube short and the stage level. Nine physicians out of ten engaged in similar examinations use their instruments in an inclined position, covering their preliminary mounts, absorbing the superfluous moisture with a bit of blotting paper, employing, too, an objective as high as the one-fourth, or perhaps higher. Now nine-tenths of all work of this description can be accomplished with a wide angled inch or two-thirds, and by keeping the tube short and the stage level there will be found, in ninety-five cases out of the hundred no necessity for covering the mount. I simply place on a clean glass slide a drop of the specimen to be examined and, without covering it at all, place the same on the stage, the spring clips being turned back out of the way, for even these are a drawback to rapid work. It will be often necessary to use re-agents, and to this end the long working distance affords every facility. With this two-thirds or the inch I am enabled, by eye-piecing, to get nice definition up, say to 200 or 250 diameters. Here we have a *practical* advantage arising from the use of a high-angled glass, and one of the greatest value. In thus being able to dispense with the use of covers a wonderful saving of time is accomplished; the objective, too, is far enough out of the way, so that it does not become clouded by the evaporation, nor injured by the fumes of the re-agents. Any desirable change in amplification that can ordinarily occur is furnished instantly by changing the eye-piece. And here let me say that the oculars should slip in and out of the tube just as easily as possible without decentering the object viewed. A tight fitting eye-piece is an abomination of the first order; any ocular of mine will drop out of the tube instantly by tipping the stand " upside down." Now the advantage of the short tube is this: You are enabled to work over a table of the ordinary height, and to view your object comfortably, at the same time the table forms a very acceptable rest for the forearm; on the other hand, by keeping the tube its standard length one must use a lower table, and the rest for the arms can no longer be obtained unless some recourse be had to blocking up by books and the like.

Now in an examination involving three or four hours' time, it is quite possible that it may be desirable to substitute in the place of the two-thirds a glass that will give higher amplifications. In this case the two-third would be removed and the one-sixth called on to take its place. Now I *have previously found out* that the one-sixth has maximum performance when worked over the thickest covers. These I have already selected and placed in a little box by themselves, one of which is carefully cleaned and placed over my object, the microscope tube pulled out to the standard length, the instrument inclined to a suitable angle, and thus the examination goes on. In extreme cases the one-tenth would be again substituted for the one-sixth. With this glass, having as it has, maximum performance at any point of its cover adjustment, there need be no particular care exercised as to the selection of the cover, further than to see that the same be thin enough.

It will be noticed that with the employment of the immersions the tube should be restored to the standard length. This is an important item, and should never be omitted when there is nice work in hand. These wide-apertured glasses are especially intended by the optician to be worked with tubes of standard lengths. The range of collar adjustment, too, is in many instances arranged conformably thereto.

In observations over urinary deposits I contrive to do a great deal of work with the one-half inch. Of the two, the two-thirds or the wide-angled inch is much the superior glass; but I value these so highly that I dread to use them over the fumes of chemicals. Hence I make the cheaper glass do all that I possibly can, and in this kind of a way the glass is really useful to me.

It will doubtless be observed that in changing from the inch or two-thirds to the one-sixth or one-tenth I make a pretty big jump. This has often occurred to me, and has led to the trial of several intermediate powers, resulting in every instance in my going back to first principles; we are still of the notion that a No. 1, four-tenths, or say three-tenths of the very highest aperture possible, would be a valuable glass in the laboratory, especially in such examinations as those of urinary deposits or histological work generally, and we hope before long to own just such a glass, which, if a success, shall not be allowed to " hide its light under a bushel."

In the examination of malignant growths, and in the study of minute pathology generally, the aforesaid programme is somewhat modified. In this line one can very often make entire and satisfactory examinations with a "medium power" glass; therefore, and principally for the sake of convenience, I have in reserve a dry one-fourth of 100°. This objective (the dry front to my one-sixth) will give me nice, clean and reliable views under amplications, say from 200 to 600 or 700 diameters. · The mechanical working of its screw collar is smooth and efficient, and the glass responds promptly to any change thereof. Hence it has worked its way into favor and use. It is not a strictly necessary objective in the laboratory, for the one-sixth will of course do all that can be expected of the one-fourth, and a great deal more besides. The one-fourth is used precisely under the circumstances stated; when there is work on hand likely to call for the shifting of objectives, the one-fourth is very likely to remain in its box. Thus it will be seen that my " working " battery of objectives is not numerically very formidable, and I may add, not very expensive.

Having thus stated my own course in the way of selecting objectives for my particular work, let us now turn our attention to a condition of things which is occurring every day, to wit: A young physician just graduated wishes to use the microscope in his (expected) practice. It is important to him that the investment in an outfit be made with reference to the strictest economy. The author has in the past, and is now, receiving many letters of this sort, to one of which he replied but a day or two ago substantially as follows:

The fundamental idea in purchasing an outfit ought to be this: To buy nothing with the view of replacing it bye-and-bye for something better of the sort, with an indefinite hope that the original article can be disposed of at no great loss. On the contrary, let every purchase be made and every detail carefully selected with the intention of avoiding any future substitution or exchange. Any departure from this fundamental law will be at the sacrifice of strict economy; therefore we reply to our correspondent in this tenor, recommending that he pur-

chase a one-inch like those furnished by Mr. Gundlach
or the Messrs. Beck; and for a high power select the
professional one-fourth of the Messrs. Spencer & Sons,
or its equivalent by any other maker.   With these two
glasses the physician can accomplish seven-tenths of any
and all work he may be professionally called on for.
Either one-inch named is a good reliable glass, giving,
when working with suitable oculars, very good shows
indeed, and infinitely better than the imported French
triplet, while the cost is but very slightly enhanced.
We recommend a one-fourth similar to the Spencer professional, because, having used one ourselves, we can
speak "by the card."   One very strong reason is, that
the adjusting collar mechanism (although moving the
front lens) is very smooth and satisfactory in its working.   A still more important reason is that these glasses
respond nicely to any movement of the collar, and they
are, too, well corrected throughout the range of their
aperture.   The purchaser is thus thrown *early* in contact with an adjusting objective, and hence by his daily
practice will he become more and more skilled in its
use.

   The author desires the reader to make a special point
of this: Many there be who, having bought adjustable
objectives of poor quality, and having discovered "*practically*" that the position of the collar adjustment with
such glasses hardly affects in any way their efficiency of
performance, not only let them alone for the future, but
settle down strong in the faith that collar adjustment
don't amount to much, and that the "handling" of an

objective sometimes read about is simply a myth. Here is another instance where the French proverb fits to a charm. The fundamental error having been committed, the evil consequences are sure to follow suit.

Now in the course of a year or two, our correspondent having meanwhile obtained some considerable knowledge of the microscope, and being in a position to make a further investment, he has simply to purchase a first class inch (or two-thirds) of 45°, the old inch will still do good service either as a "hack" lens for his rough preliminary work, or it can be made to do yeoman's duty as a condenser, as has already been referred to. It will, in very fact, be just as much needed as before; there will be no thought or occasion for its sale or exchange.

It will probably happen, too, that in a little while the one-fourth will be supplemented by the addition of a one-sixth or one-tenth of high balsam angle. In this case the great advantage derived from the previous use of the one-fourth will at once be made manifest. Our correspondent will be enabled to work the new glass with tolerable satisfaction at sight, improving daily as he continues its use; and if so be that he can get a few words of instruction from some acknowledged expert, they will be readily understood and appreciated. And here, again, note that the one-fourth does not even now become a superannuated, worn out objective, to be sold at the first chance to the highest bidder. On the other hand, it will continue to be used generally for a considerable time while experience with the last purchase is being gained. In short, it will remain " a good thing to have in the house " until some genius like Spencer or Tolles shall generally upset all our arrangements by some master stroke of advancement. There is no defence possible to provide against such an emergency.

We desire here to say that our remarks as to objectives also apply as to the purchase of stands. The man who purchases a stand "just for a year or two," as very many have done, and are still doing, does so at a sacrifice of true economy.

But we have correspondents of another class. Says one, " the inch you name costs $7.00; the one-fourth inch, $20.00, making $27.00; now I can't, for stand and objectives, spend but $45.00, or at the very most, $50.00; and then there is a sub-stage arrangement recommended; also, an extra eye-piece or two. What shall I do? This *is* all the money I have got, and I have to take the clothes off my back to expend this much."

Now to all such, and there are many of them, we reply, make a " *virtue of necessity.*" The situation *is* unfortunate, but, as you say, it cannot be helped. If, by waiting a short time, things can be improved, then you had better postpone your purchase. If to the contrary, then invest as best you may under the circumstances. Keep this in mind: Of the two evils, *sacrifice the stand rather than the objective;* the latter MUST be maintained up to the standard given. You might possibly substitute, in place of the one-fourth, the best three-tenths obtainable, non-adjustable; but there

would even then be but a few dollars " saved," with the
fact staring you in the face that you would lose all your
practice with an adjustable glass.   This you *must* have,
and from the very start.

The author thus dwells on this matter because he feels
that it is one of deep interest to many.   It is one, also,
that he has given particular attention to, that he might
be able to advise others intelligently; and he is not
without reason to hope that the information he has thus
tried to convey in the simplest possible language will
be to many worth more than the cost of his little book.
He is aware, that in writing the last page or two, he
has been practically repeating the tenor of what has
before been written; but nevertheless he is strong in
the faith that more than one of his readers will have no
fault to find with him on this score.   To those about
purchasing an outfit of objectives he would further say
that the advice here given is precisely what he has given
to scores of enquirers during the past six years, every
one of whom have expressed their satisfaction in a manner not to be mistaken.   Let the reader, however, be
reminded that we live in an age of progress, and that
it is quite probable that the instructions offered here at
this date may at no distant day require essential modifications.   Doubtless improvement will continue to follow
improvement in the future as well as in the past; the
genius of our American opticians is such that anything
in the way of progress successfully accomplished seems
to render them more restless than before, and in view
of this, to use a nautical phrase, we advise the reader

to keep well " an eye to windward." And it may be here we can render *quasi* assistance, thus:

Judging the future from the past, it seems more 'than likely that our opticians will at no distant day produce one-fourth, one-fifth or one-sixth quite equal in every respect to the one-tenth of to-day, and possibly with a greater working distance. The author, within the last twenty-four hours, has received reliable testimony affirming that young Spencer has succeeded in making a wet and dry one-sixth fully up to the performance of the one-tenth, of which mention has before been made in these pages. This one-sixth we have not seen, but are informed that it is now in the hands of an eminent microscopist, who will exhibit it at the Paris exposition.

Again it may be possible that the optician will succeed in enlarging the aperture of the low powers. Heretofore 45° has been regarded as the limit of angle for the inch; with our present knowledge it appears almost an impossibility to extend the present limit much beyond the figure named, unless, indeed, the calibre of the objective be increased. It will be advisable, nevertheless, to bear in mind that in the late march of advancement of the American objective, several desirable points, formerly declared " impossible " have been mastered by the optician, and are to-day "*un fait accompli.*" Now if it shall be so that the optician succeed in producing a one-fourth equaling in angle and performance the present work of the one-tenth, or that the one-inch shall in the future rival the work of the two-thirds of

to-day, then will the superiority of such glasses be
demonstrated, and for reasons already given, and we
are disposed to look in this direction for future improvements. Let the reader also remember that the item
of "working distance" must not be lost sight of. For
.instance, if, in the extension of the one-fourth to the
capacity of the present one-tenth, the working distance
of the former should be decreased to that of the latter;
then the real gain would be comparatively slight,
amounting practically to a saving of a few dollars in
the cost of the glass. The true problem is to gain
working force and efficiency without sacrifice of working distance. We recognize what has been accomplished, and the recognition is suggestive of further
advances in the same direction.

SELECTION OF COVERING GLASS.

This can be readily obtained of the dealers either in
sheets or ready cut into squares or circles, and of any
desirable thickness. As to the mere form of the cover
used, that is a matter of fancy; but the *thickness* ought
to correspond to the working of the objective. We
try to confine ourselves to three thicknesses of covering glass, namely, one-seventieth, one-one-hundred-and-
twentieth and one-two-hundredth of an inch. These
may respectively be denominated as thick, medium and
thin. It is a matter of the first importance that those
working first-class objectives should be well posted
as to the thickness of cover employed, and yet this tell-
ing point has been utterly lost sight of in the books.

For example: By knowing the thickness of the cover, one is enabled to approximately adjust the objective at sight, and thus save time. We have thousands of mounted objects in our cabinets, and .every cover has been measured with all the accuracy obtainable. Those who have long had their attention called to this item can, by dint of practice thus obtained, tell closely the thickness of the cover by simply *feeling* it, and this, let me assure the novice, is an accomplishment worth having. Said the veteran microscopist of New York, Rev. Dr. Armstrong, to the author, not long since, "It's astonishing how fast you work. You seem to be in perfect harmony with the objective." Now the secret of the fast work noticed by the doctor lay in the fact that I knew the thickness of all my covers, and was thus enabled to place the collar of the object-glass very closely in position at the very start; and not only this; I was, for the same reason, posted as to the use of water or glycerine. Now, to take a case from practice: Suppose I desired to examine a brand new mount. Let it be a difficult diatom this time. First, I run my finger over the cover and instantly discover that it is a thin one, say about like those used on the Moller plates. Now, if I elect to use the one-sixth objective, I know that this cover is too thin for water immersion ; hence glycerine is chosen. I know, too, that over such a cover, and with the glycerine intermedium, the objective will "correct" some three or four divisions from closed; therefore the collar is at once placed near such position. Now, on looking through the tube at the object in position and focussed, suppose I do not get as good views as I had reason to expect, then *I let the collar stand as it was* and change the illumination until things are approximately as desired; this done, a slight turn of the collar adjustment will insure the maximum working of the objective. Now just contrast this with the usual " modus." Eight operators out of ten would have at once twisted 'round the collar hap hazard like, by " rule of thumb," probably wasting plenty of time, and, more unfortunately still, condemning a really good objective, and one that would have, with the proper manipulations, given charming displays.

Now, in the instance quoted, we took the mount as we found it, and "handled" it as best we could with the least loss of time, the only means at our command being the choice between glycerine and water as the immersion medium. Now suppose the mount was one prepared with especial reference to a one–twenty-fifth or a one–fiftieth, and having an extremely thin cover, say one–four-hundredth of an inch thick. Under these circumstances it would have been imperative, to suit the work of the one-sixth, that this thin cover be supplemented with another and a thicker one, making *optical* contact with water or soft balsam. This *is* an inconvenience which, in the case alluded to, cannot be well avoided; but we are nevertheless taught the propriety of suitably covering such mounts as may be prepared in our usual line of practice. All this seems plain enough, without further demonstration. Let the tyro then cover his mounts as far as possible to suit the objectives under which he expects to show them, and let him, too, early learn to discriminate as to the differences in thickness of covering glass.

A handy instrument for measuring the thickness of thin glass is a little pocket micrometer gauge manufactured by the Brown & Sharp Company, of Providence, R. I. Although not expressly made for the purpose, it is as convenient and accurate as any other we have met, while its cost is very moderate.* We make it a plan that when a quantity of new covers are first in hand to pass them over to the pupil, to be by him measured with the instrument aforesaid, and assorted according to their thickness, the various styles being kept in little boxes by themselves, and duly labeled. In this way the student acquires the needful practice, and the requirements of the laboratory are met at the same time. If so be that no instrument of the kind is at hand, or that the observer feels that he cannot afford the necessary outlay, the usual fine adjustment of the stand can be made to do tolerably effectual duty, thus: Obtain a cover, of which the thickness is known, and with a fine pen dipped in thick India ink, make a mark on one surface near the centre; turn the cover over and make another similar mark nearly, but not exactly opposite the first. Now place the same under the microscope, focus on the mark on the under side of the cover, and then, by the fine wheel of the fine adjustment, focus again on the mark on the upper surface, noting the revolutions or parts

---

* As we have before remarked, the fine adjustment of the Acme can be used as a micrometer.

thereof traversed by the wheel. By this simple procedure a gauge is established from which other covers can
be easily measured. Some stands have the wheel of the
fine adjustment graduated, which will be found especially
convenient. In other cases the observer can make his
own graduations on a little circle of paper and adapt
the same to the wheel, without any great drain on his
mechanical skill.

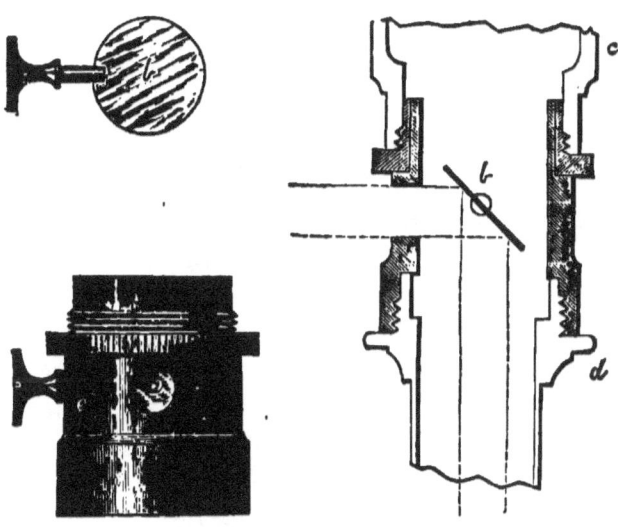

BECK'S VERTICAL ILLUMINATOR.

This instrument consists of a brass tube, either end
being fitted with the "society screw," so that it can be
attached to the back of the objective; the object-glass
with the illuminator thus attached is screwed to the nose
piece of the stand. The tube has an adapter of its own,
so that it can be revolved around the optical axis of the

microscope. In this tube is placed a circular glass disk (one of the ordinary circular covers used in mounting object slides). This disk is supported in position by a horizontal pin, to which it is fastened with a bit of cement, the pin passing to the outside, and terminating with a little knob, by which the disk can be revolved at will. The tube is pierced laterally with an aperture for the admission of light. The light from an ordinary hand lamp passing through the aperture is received by the little glass disk, and, by turning the outside knob properly, is made to pass vertically downwards to the rear of the objective, and by it in turn concentrated on the object to be examined; no sub-stage or other illumination being used. This is the instrument as made by the Messrs. Beck, and is in fact but a modification of a similar instrument invented by Prof. H. L. Smith, of Geneva, N. Y. Prof. Smith's device, however, is furnished with an interior metal reflector in place of the glass disk of the instruments of the Messrs. Beck.

We had used the illuminator but a very short time when we discovered that the definition of the objective was very much improved by shutting off the area of the lateral aperture, thus allowing less light to enter. We also found that the actual amount of light needed depended on the objective employed. In most instances the higher the angle of the object-glass, the smaller should be the area of the lateral aperture.

About the time we had arrived at the above fact, the Hon. P. H. Watson happened to call in to spend an hour or two " over the tube." The conversation turning on the action of the Beck illuminator, we coupled the instrument to my Tolles one-tenth and worked them over Mr. Watson's Nobert test-plate, and in a very short time succeeded in getting a most charming display of the 19th band, and while the latter aperture was nearly closed by interposing the circular edge of a large bull's eye condenser which happened to be at hand on the table; Mr. Watson and myself were both delighted with the exquisitely beautiful display of this so-called difficult object—this 19th band. Other severe tests were also taken in hand and resolved. Among others we had a glorious show of podura under amplification of some 4,000 diameters. It was demonstrated, too, that the very best resolutions were only obtained when more or less of the lateral aperture was closed by interposing the circular rim of the condenser, the clear aperture left being in the form of a crescent. Subsequently Mr. Watson devised the following described attachment, which answers the purpose fully:

The lateral aperture of the instrument is somewhat enlarged and a hollow plug inserted therein, the opening in this plug being the same calibre as the original aperture. The plug is somewhat tapering, and fits the opening "spring tight." It also projects outward slightly beyond the circumference of the main tube. To the outer end of the plug is fitted a little shutter, while a narrow slit, about one-one-hundred-and-fiftieth of an inch wide, is pierced through the shutter; the whole so arranged that the opening the plug can be partly or wholly closed, or the little slit used by itself; and when low powers are

employed, or objectives of narrow apertures used, the
entire attachment can be removed if desired. This
illuminator, as originally designed, was intended for
use with dry objectives and with moderate magnifications; the idea of using it with immersion glasses of
high angles and under high powers, originated with
George W. Morehouse, Esq., of Wayland, New York,
who is well known as an expert and accomplished microscopist. The special advantages obtained by the
immersion system in the case in hand are too apparent
to need further mention. It will be noticed, too, that
the Beck is thus made to do duty—like the Wenham
reflex—in a way quite foreign to the purpose of its
original inventor.

The advantages derived by the use of the instrument
are: First, we are thus enabled to view objects by the
aid of reflected light; the so-called "opaque illumination," under the highest amplifications and (if the proper
objective be employed) with superb definition, as is
attested by its unequaled work over the Nobert 19th
band; and by the employment of artificial light, a common kerosene hand-lamp being all that is required. In
this respect the instrument stands alone and inimitable.
Second, the views given are surface markings only.
There is no "penetration" here. The focus must be
most accurately drawn, and on the surface of the object.
The slightest deviation therefrom is instant and total
defeat. Thus we are enabled to *locate* structure, at
times a most valuable assistance to the observer. Third,
by a slight change in the position of the lamp, the mir-

ror being at the same time brought into play, the illumination may be almost instantly changed to that of transmitted light, or from thence back to reflected light again. This cross-questioning under two methods of illumination is often of great advantage.

Per contra: The vertical illuminator has but one drawback, and that is rather a serious one, to wit: Objects to be displayed under it must be mounted dry, and also contact the cover. Hence it will be seen that a large portion of histological, pathological, as well as other permanent mountings, are excluded from use. Many of these can be temporarily prepared for study, and it is hoped that the attention of observers will be enlisted in this direction. Admitting the serious character of the objection named, let us bear in mind that there remain countless fields of research wherein the " Beck" will certainly prove an instrument of the greatest value.

The tyro will find the Beck illuminator a difficult instrument to use, and his first attempts will probably result in failure. Nor is it an easy task to give any instruction in writing that will be of much aid. The instrument not being a costly one, and likely to be generally used when it shall become better known, we will try and furnish a few hints that perchance may prove of value to the beginner.

The novice will do well, at his first attempts, to select a dry mount, one that he is perfectly familiar with, and preferably scales from the lepidoptera; a dry mount of podura will answer very well, indeed. Select, also, the highest angled objective, and with this examine the mount in the ordinary way and get a tolerable correction for the glass. Next, removing the object-glass from the stand, couple it to the illuminator and screw the whole to the nose-piece of the stand, using transmitted light, as usual. Focus the objective. Now hunt through the slide; among the numerous scales will probably be found one or two which, in order to bring in focus, the objective will require to be withdrawn from the cover slightly. In such a case the chances are that that particular scale is nearer the cover, and if in good condition may be selected for further operations. Next, bring the lamp (a flat wicked one) towards the observer, revolving the tube of the illuminator so that the lateral aperture shall be in proper position to receive the light from the lamp, the latter being about seven or eight inches distant, and the flame about the same height as the aperture of the illuminator. Now grasp the little knob connected with the interior glass disk and turn it so that light shall be reflected to the rear of the objective; at the same time, and looking through the tube as you catch the first glimpse of light, revolve simultaneously the main tube and also the little knob carrying the glass disk, the object being to secure as great an amount of light as possible. A little manipulation of this kind ought to result in illuminating the object with a horizontal (or nearly so) band of light. The next step will be by a slight movement of the lamp, keeping its edge *exactly* towards the aperture, to endeavor to make the band of light crossing the field as *narrow* as possible, and the outlines of the band clear and distinct. By this time the operator will have probably discovered that a slight rotation of the main tube will separate the horizontal band into two parts, or, as some of my pupils express it, " two tongues." The best position is when these are made to coalesce as completely as possible. It is also probable that in the attempts thus far made that the image of the scale has been well seen. When this occurs it should be at once focussed. The next procedure is to correct the objective; the correction obtained by transmitted light will not suffice for the purpose in hand. It will be noticed that as the glass is made to approach the correct adjustment, the horizontal band of light will be correspondingly improved. So true is this, that one might almost be governed thereby in the adjustment of the objective. Having got thus far along, and without any serious mishap, it will be easy, by closing the shutter, to admit the precise and most favorable amount of light, and also to try the effect of sundry *very slight* changes in the position of the main tube, glass disk and lamp. Very beautiful resolutions are sometimes obtained by bringing the lamp within five or six inches and interposing the bull's eye condenser, flat side to the lamp, in which case the shutter must be further closed. It will happen also, occasionally, that the best exhibition of striæ on very difficult objects, such as extremely close *frustulia saxonicas*, is when the striæ are placed at right angles to the horizontal band of light.

Now, should the manipulator meet with tolerable success, and get good shows of the lepidoptera, I recommend that he practice for some half-dozen sittings over the same mount. By this he will get a certain familiarity with the instrument which will be of great value to him. The chances are, too, that he will wonderfully improve in the manipulations, and in this alone will be well enough rewarded for his pains. Furthermore, he will learn what nice effects can be had by the slightest changes in the position of the tube, glass disk, shutter or lamp, and the necessity for the closest focussing will be taught practically.

Having thus got tolerably acquainted with the instrument, let the operator try changing from reflected to transmitted light, and *vice versa*, which is accomplished merely by moving the lamp back to its first position, employing the mirror just as was recommended at the start. A little practice of this nature, working alternately by transmitted and reflected light, will soon accustom the observer to the situation, and enable him to make the change in illumination almost instantly.

A fact worth knowing is this: The combined length of the illuminator and objective will, if the objective selected be one of the extremely long models, defeat its use on some of the smaller stands. Thus we find that we cannot use the illuminator with a Tolles one-sixth or one-tenth on the little "Histological" of Mr. Zentmayer, there not being sufficient room between the "jacket" in which the body tube slides and the stage to receive the illuminator and objective when coupled together. We are therefore, when essaying the use of the Beck, compelled to fall back on a larger and heavier stand.*

BULL'S EYE CONDENSER.

This instrument, so well known, and accompanying almost every stand as sold, we make nearly constant use of. It should be simply a large plano-convex lens, say from two to three inches in diameter, and fitted with universal mountings. It has been too general a rule with the microscope makers to "adapt," the size of the condenser to that of the stand. Thus, if a certain maker furnish five or six different sizes of stands, he will be pretty sure to have as many sizes of condensers to accompany the same. But let the reader insist that, however small may be the model of the stand selected, the condenser be at least two inches in diameter, and that even three inches will be found none too large. Any of our opticians can furnish the instrument made to order. We have seen and worked with several made by Zentmayer to accompany his large and intermediate stands, and also the large model furnished with the Messrs. Beck, all of which were effective instruments.

WORKING WITH LOW POWERS.

It has been our primary intention throughout this work to avoid repetition of instructions, hints or suggestions, such as may be found in the various textbooks, while interspersed among the pages already writ-

---

* The construction of the "Acme" admits the use of the Beck illuminator.

15 Microscopy.

ten the reader may perhaps recognize information as to every day work which, perchance, he may turn to some account. There remains therefore but little for us to add under the above caption. Without hesitating to repeat in a more condensed form the same ideas which have heretofore been scattered through our pages, we proceed to give other methods of working with the lower powers.

First, we use and recommend a stand fitted with a swinging sub-stage, preference being given to the one that will allow the mirror to rise above the level of the stage. Stands there are which although allowing a limited swing, do not afford the extreme range of motion, while scores there be fitted with stationary sub-stages and fittings. Of the last two named, that with the limited range is infinitely to be preferred. One of the principal advantages arising from swinging the mirror above the stage is that we are enabled thus to condense either sun or artificial light on the top of the object, and it is possible to accomplish this otherwise than by the swing, by simply attaching a mirror to a separate adjustable stand of its own. In whatever arrangement which may be selected, let it be *imperative* that the mirror be attached directly to the swinging arm, and at the proper focal distance, and that the centre of rotation coincide with the plane of the object on the stage. Any adjustable and intermediate joints between the mirror and the swinging arm, allowing the former to be placed *out* of its focal position, is, in the opinion of the author, an intolerable nuisance, and one not to be submitted to under any conditions short of sheer necessity. Now of the two styles of stands, the one with the full swing and the other with the partial swing only, premising that both mirrors rotate in the plane of the object, there will not be a great deal of choice for general work. Nevertheless it costs no more to manufacture the one than the other, and hence the perfect stand becomes no more costly than one that is to a certain extent imperfect. Therefore we urge, why not procure the best, provided there are not especial and governing conditions patent to the purchaser. The substance of all this has been placed before the reader, and in again calling his attention, it is hoped that the repetition will gather force.

Now all the information we have to offer as to working with the low powers will refer entirely to the use of the low-angled sub-stage condenser conjointly with such a stand as we have recommended. As has been before remarked, the introduction of the swing has worked a radical change in our ideas as to the value of the condenser, and the instrument is now with us a constant fixture, and in daily use.

With the use of the condenser we have at command a greater amount of light, but this is not the special advantage derived from its use; and then again it is easy to make the bull's eye do duty in its place. Of the two accessories, the bull's eye is the more simple and convenient to use; nor will objects occupying the entire or greater portion of the field when illuminated with the condenser appear more brilliant and enticing than without it. The grand advantage obtainable with the condenser is, that by a careful manipulation of the apex of the small cone of light we are often able to get just the right illumination on certain details of our object, while the field generally is kept in partial light only, " toned down ". as it were. This effect may be produced in two ways: First, by raising or lowering the condenser; and, secondly, by swinging the sub-stage laterally to a point but just within the aperture of the objective employed. We sometimes use the one plan and sometimes the other; and then, again, we often use a mixture of both, and, as a rule, the better illumination will be attained when the condenser is somewhat within, or without the focal point.

Now let the novice understand that in thus employing the condenser there is no attempt to get " pretty displays " of the object. On the other hand, *the* primary object is to obtain cool and reliable definition of structture. Those who thus use it will, I think, be pleased with the results attained. It is somewhat more " bother " than taking the light directly from the mirror, or with the usual intervention of the diaphragm. Notwithstanding this, one soon becomes accustomed and addicted to its use. The swinging stage, too, is a valuable adjunct to the old diaphragm, and with this latter instrument superior effects can be had working on the same general plan as with the condenser, and using a small aperture to the diaphragm. Should the light be too weak it can be assisted by interposing the bull's eye, or light can be shut off by depressing the diaphragm.

For ordinary work by daylight we use the plain mirror in conjunction with the condenser; and let me here again insist that the latter be such as has already been described, *i. e.*, of low aperture, admitting but a narrow cone of light. In very dark days the light can be reinforced by employing the concave mirror, a plan not generally recommended, but nevertheless quite practicable.

For reasons before given, we greatly prefer, for regular right straight along daily work, to use the stand with the tube short, keeping, for the most part, the tube vertical, dispensing, too, with the stage clips, and simply laying the preliminary mount thereon, shoving it about in every direction required, by the fingers. Those who have been accustomed to confine the slide under the clips, as is generally done, will not, on the first trial, be likely to endorse my practice. Let me, then, to all such, especially recommend it. By thus allowing the slide to rest by its own gravity alone, one soon acquires a delicacy of finger manipulation that is of very *general* value; while, *per contra*, the clips are a real hindrance to fast work. Those who have the patience to practice without them for one solitary week will not be likely to get back into the old rut.

The height of the work table should vary with that of the observer; such a table as one would naturally elect to write on will be about the correct height for microscope work. This, if the low angled condenser is to be generally employed, may be placed at a considerable distance from the window, but care should be taken that the light comes from the left. If there be other glazed openings in the room liable to form cross-lights, they should be closed by shutter or curtain. The table selected should be solid and heavy. Any and all of the little light affairs in cherry and mahogany offered at the furniture stores are totally unfit for a microscope table; but the crowning nuisance of the lot is the revolving table, made expressly for the microscopist, and sold at outrageous prices. Castors, also, are to be rejected. In short, anything that detracts from the solidity or rigidity of the work-table is to be eschewed. What the observer wants is a firm support for his stand, free from shake or tremble—one that he can lean against freely when weary, and one that he may even run against accidentally without inaugurating any serious calamity. A couple of drawers placed in the front are a convienience for storage of accessories, etc., and these, being partly opened, form convenient rest at times for the forearm. The table should be sufficiently large and roomy. Three by four feet is none too large.

THE SPENCER ONE-INCH OF 50°—BROAD-GUAGE OBJECTIVES, ETC.

In the spring of 1878 we received from the Messrs. Spencers an inch objective of 50° aperture. This glass was made expressly to our own order.

Our purpose in ordering the above glass was to determine whether it was possible for the Messrs. Spencers to furnish an inch objective having angle and definition equal to their celebrated two-thirds.

With the new inch our experience has thus far been

necessarily limited; we have, however, instituted a few comparative tests which we now present to our readers.

The working distance of the two-thirds worked with one-fourth inch solid ocular and standard ten-inch tubes is twenty-five-one-hundredths of an inch.

The working distance of the new inch with same ocular, worked with ten-inch tube, is thirteen-one-hundredths of an inch; when worked with the same eye-piece and the short tube of the "Histological," the working distance is increased to eighteen-one-hundredths of an inch.

In general, the definition of the inch is superior to that of the two-thirds. The very best definition of the latter is obtained by the use of the full length tube and one-fourth inch ocular, while the highest definition of the inch is reached by the use of the *short* tube and same ocular.

The inch, worked over the balsamed Moller plate, gives readily nice shows of the transverse striæ of *P. Balticum*, and with a little management both sets of lines are brought out. On the *G. Marina*, immediately preceding the *Balticum*, and really the more difficult shell, the inch gave me a very nice stand of the transverse striæ from end to end of the shell, its performance over this diatom being manifestly superior to that of the two-thirds.

On dry mounted shells of the *P. Balticum* the two sets of lines are very well displayed by the inch, as are also the transverse striæ of dry *angulatum*, a favorable frustule being selected.

Over the Nobert test plate the inch gave me the 9th band, = 56 × lines in .001, Eng. inch.

Working the two-thirds with the one-half inch ocular and full length tube, and the inch with one-fourth inch ocular and *short* tube, the amplifications were about equal; the inch in every instance affording the better definition.

According, then, to the inch manifest superiority in point of defining power, its shorter working distance, which in its most favorable aspect is 33 per cent. less than that of the two-thirds, must receive due consideration, this decrease, too, of working distance will be accompanied by some loss of "penetrating power."

A much mort important point, however, to the author was to ascertain if the focal distance of the inch was sufficient to allow its being worked over wet and uncovered mounts without clouding from the evaporation. To test this, we used the glass an entire evening in regular routine work over urinary deposits, and without experiencing the least trouble from the source named; while on the other hand the general behavoir of the objective while thus engaged in practical work was most satisfactory.

The reader is reminded that in comparing working distances of long focus glasses, the same percentage of weight does not obtain as would occur in comparisons of object-glasses of nominally short focal lengths, and giving high amplifications.

A word or two concerning the standard " society screw:"

After experimenting with the wide-angled inch, as above related, we wrote to the Messrs. Spencers, asking them if it was possible either to still further extend the angle of the inch, or, maintaining the same aperture, to increase its working distance. In reply, the elder Spencer informed me that it would be difficult to further increase the aperture or the working distance unless the diameter of the objective should also be enlarged.

From this it would seem desirable to increase the diameter of our low-power object-glasses, and this in turn necessitates a change in the construction of the microscope stand. Mr. Bullock's large stand, as also the *Acme*, are especially arranged to accommodate these " broadgauge" low-power objectives.

An one inch objective of large calibre is now in process of construction for the author. The optician hopes to endow it with an aperture of at least 62°, maintaining a working distance of one-eighth of an inch.

It is but a simple act of justice to say that the idea conveyed to me by Mr. Spencer had already occurred to Dr. W. W. Butterfield, of Indianapolis, Ind. While in attendance at the congress of microscopists held in Indianapolis in 1878, Dr. Butterfield showed me a broad-gauge four-inch made to his order by London opticians. He had also a stand then in process of construction, and designed for the use of this class of object-glasses. This, however, at the date mentioned, had not arrived; consequently there were no accommodations for an examination of the objective. Should my own glass be completed before these sheets go to press, some account of its performance may herein be expected. Meanwhile, there seems to be no good reason why, with the increase of calibre of our low-power objectives, there should not also obtain those advantages due to aperture and working distance.

CHAPTER VII.

For this class of work we have in the preceding pages unhesitatingly expressed our preference in favor of objectives of the widest aperture. Such are the instruments we ourselves use daily, and can confidently recommend to all who may be desirous of working with the best instrumentation obtainable. Therefore, whatever we may have to offer in the way of instructions suggested by the above heading will be solely applicable to the class of objectives generally known as " wide-angled," to which we have given in the past, and propose yet to give in the future, a large amount of careful study and attention. And first of all it becomes necessary to disabuse the mind of the student of some of the popular fallacies which have found outlet and circulation through the medium of the microscopical periodical literature of the present and past few years. These, as will be discovered by the attentive observer, are paroxysmal in their nature; in fact are veritable " chateaux en espagne," at once inconsistent in detail, and roundly absurd when contemplated as an entirety. Thus it occurs that at one moment the student is taught that wide-angled glasses are extremely inconvenient; that great attention has to be bestowed on the adjustment and illumination, etc.; while on the other hand, another " au-

thority" subsequently insists that the so-called "handling" (?) supposed to be necessary to the use of the
wide apertures is simply a myth—a downright farce;
and that any one possessing a fair quota of intelligence
can easily acquire all that is to be acquired in the working of an adjustable glass. Nor need one hunt long or
dig very deep to find other "authorities" teaching that
all this handling" "*although essential to the optician*,"
is no manner of use to the practical observer, unless he
has so far degenerated as to aspire to the distinction of
being simply a "handler" and a "fighter."

Let the author, then, and in view of the situation as
presented, inform those proposing to study the microscope
with the intention of becoming in due time accomplished
observers, that there is no "royal road" to success; that
to become even so sufficiently expert as to enable one
to *follow* out (leaving original work out of consideration) the investigations already made and published by
eminent microscopists, will require quite as much effort
and study as would be called for in graduating from any
college in the United States. The curriculum is a broad
one in its very nature, involving a thorough knowledge
of instrumentation, and when by means thereof we are
enabled to *see* well, it then becomes a positive necessity
to *judge* well of what is seen, and this in turn can only
be well accomplished by those having eyes well trained
to the work in hand. There is, moreover, work for the
brain outside of the functions of the optic nerve.

In the micrographic dictionary, by Messrs. Griffith &
Henfrey, I find, (page 11,) the following: "Above all,

however, it must never be forgotten that microscopic investigations require more time and patience than perhaps any others, even in regard to the determination of simple facts of structure and qualitative composition; and although it is not very uncommon to hear those engaged in them sneered at as wasting their time over a very simple plaything, *this may be regarded as arising from one of those prejudices which will exist so long as people will venture to express opinions upon matters with which they are unacquainted,* and which are beyond their comprehension."

The above quotation is well worth reprinting on its intrinsic merits, and it may be that we shall find some especial use for it bye-and-bye. Meanwhile, this matter of eye training calls for a word or two, for, among the accomplishments that go to make the first class observer, this education of the eye is generally supposed to be quite as much a myth as the capacity to "handle" an object glass. Let us take an item or two from our personal experience:

Not long since the author had the honor to address a select party of gentlemen at the parlors of a private residence. In the course of his remarks the matter of "eye-training" was brought prominently forward and its usefulness urgently insisted on. Subsequent to the close thereof a gentlemen present stepped up to him, saying, "I want a little talk with you about that eye business, which you seem to regard as a *sine qua non.* I do not see that thing as you do. You and I are about the same age. We have both of us necessarily been

using our eyes constantly all our days. If I look across
the street and see a house, why, so can you; and thus
we have been respectively seeing houses as well as other
things all through life. In short, our eyes have been
constantly at work, and have thus been as constantly -
trained. It may be that you have abused yours by over
work with the microscope—if that be the case, I reckon
mine have the best of it. At all events, I can't see
how you can establish any individual superiority as to
vision."

At this moment Prof. Huber had seated himself at the
piano and was entertaining the company by his superb
renderings from classical authors. In reply, I said:
what you affirm as to the eye must be similarly true of
the hands. Prof. Huber and yourself are apparently of
the same age, and both of you have been using your
hands "all your days." Whence comes that lightning
rapidity of action; that wondrous delicacy of touch?
Think you that the professor has abused his muscles by
over work with his five finger exercises, or that you
have any claim to digital superiority? But to confine
the case strictly to the eye alone; how comes it that the
mariner can, not only detect a " sail " near the distant
horizon, but can also state with accuracy whether it be
a ship, brig, or schooner, and the direction the " sail "
may be pursuing, to all of which the passengers present
will be totally blind?

Nor need we " go to sea " to find instances illustrative
of the issue in hand. The eye of the artist recognizes,
perforce of his experience with the esthetics of nature,

beauties, to which that of the shepherd boy is innocently enough a stranger, and of the two, permit me to inquire which would make the better microscopist?

As far as our auditor is concerned, we rest the issue on its merits.

A few years ago we purchased for a gentleman well known in microscope circles, a wide apertured objective. The party was no novice, but on the contrary a real hard and close worker with the instrument; and furthermore, the gentleman had formerly filled the chair of microscopy in one of our most honored colleges. After working with this glass for about one year, he applied to me for instruction in the use of this instrument, proposing to spend his vacation with me, and for this purpose the author was delighted with the proposal and the arrangement was consummated by unanimous consent. Now the main point actuating my honored pupil was this: He had used his glass considerably—enough to discover that there were conditions involved that he could not control as he desired. Sometimes he could see better than at others; sometimes the glass would work good naturedly, and then again, at others, it wouldn't work well. In the course of instruction which followed, a slide of *navicula rhomboides* was selected (this, by the way, was a diatom) and placed on the stand for examination with the objective named. In point of "difficulty" these shells would have been regarded as average specimens. We then took some little time to explain as well as we could the behavior of the object glass when in and out of adjustment, as exhibited by the object selected. Our reason for making at this time the particular selection was this: When the objective was in perfect adjustment, the striæ were admirably well seen when the inch ocular was employed, but were assumed to be invisible under the two inch. Next, the pupil was required to adjust the object glass using the two-inch ocular only, acting, of course, under the general instructions he had received. This done he was to apply the higher eye-piece and learn of his success, practically. Our friend went at his task manfully, and fought that slide of diatoms three hours or more daily for more than a full week, constantly improving in its manipulations. Then it occurred that, getting somewhat weary of his protracted efforts over one and the same slide, he began to beg for a change to do something else. Nevertheless, he was put off time after time, until nearly at the close of a long evening's work he jumped up from his table, and running towards me, his eyes beaming with joy, exclaimed, " I have it! I see it! It's all plain sailing now!" " Well," said I, "what is it?" He replied, " I can see any shell on that slide, I care not how small nor how close the striæ, and as well with the two-inch ocular as with any other; and more than that, I can put the correction collar right on the dot without humming or hawing, and do it every time." To this we responded: " You have now solved your problem, and are ready to tackle another mount at our next sitting.

Now, reader, here is a practical case in point: My pupil had not only been improving himself in the "handling" of the object glass, but all this time his eye was becoming educated. At the commencement it would have been an utter impossibility for him to have recognized the striæ of the rhomboides with but the two-inch eye-piece, nevertheless the writer saw them splendidly, and with that exquisite definition which pertains only to the work of these high-angled glasses. Had I told my pupil this at the start, he would probably have received the assertion in a becoming manner, meanwhile entertaining some " first class doubts " under his sleeve;" full fledged doubts, too, and simply waiting for a fine day to fly.

Place the microscope in the hands of the shepherd boy; its total defeat is established. Adjust, if you will, the objective with the utmost nicety, and arrange the illumination to perfection, and total defeat still reigns triumphant. He neither has the capacity of seeing well nor of judging well of what may be seen, nevertheless, he can honestly and innocently look you squarely in the face and assert that he has as good a pair of eyes " as the next man."

We have thus taken some little pains, and hope not without profit to the reader, to establish a fact well known to all who are expert in the use of the microscope. If our views are correct, it obtains that no inconsiderable amount of time and patient care and study are each individual elements in the outlay necessary in the effort to become an accomplished observer, and if this is to be considered in the light of an evil, then let it be remembered that " there are no evils unless attended with some corresponding good." And fortunate it is, in the case before us, there is a fascination accompanying the intelligent use of the microscope knowing neither limit nor bounds, and the task of becoming well acquainted with the use of the instrument is merely a labor of love. There is another aspect of the matter which deserves a word or two, to wit: The popular idea with many is, that if there be a certain amount of eye training essential in the use of the "high powers," this has no application to those who use the lower amplifications. Says one, "you are all right about your eye education when there is such nice work as showing the 19th band in hand; but then you see most of my work is done with the inch, and that's quite another thing." Now there is just as much error here, but it is of a less serious character. The truth is, the expert can see more with the inch, and by the "expert" I mean (for the time being) those referred to who are able to display handsomely such tests as the 19th band. Every day experience with pupils in the laboratory demonstrate this fact pointedly. Thus: At the commencement of his practice, the novice is quite content with the meanest French triplet the premises afford, but in less than a month he will hold it in perfect contempt, and his subsequent progress will to a great extent be properly measured or indexed by the constantly increasing capacity to handle even these non-adjusting glasses.

In thus insisting on the necessity of the proper education of the eye, let us look for a moment to its important bearing on a particular class of observations.

We intended to refer to this bye-and-bye, but it finds an appropriate place right here. We may, however, refer to it at some future time. I allude to that class of work intimately connected with the use of object glasses of wide apertures, and over exceedingly difficult lined structures; for instance, the display of the striæ of *amphipleura pellucida*. But let the reader remember that this *class* of work is not confined to the study of diatoms.

To illustrate what we have in hand, reference is made to the following sketch: Let C-D and G-H be a sectional view of some " difficult " diatom, such as *amphi‧ pleura pellucida* or the like, the short lines 1, 2, 3, 4, etc., representing in section the elevations of the striæ; M and N being views " in plan " as seen conditionally in the microscope. Let us first consider the effect of illuminating C-D with direct central light, as indicated by the line A-B. The effect will be as shown in plan at N, to wit: There will appear but a series of exceedingly fine lines; so fine (mark the words, *not* necessarily *close*) that it will be impossible to see them with any glass extant. Recourse must then be had to " oblique illumination." Now let C-D, as duplicated at G-H, be illuminated by the oblique beam E-F. The effect of this is shown in section at 1', 2', 3', etc., and in plan at K, where we have the view as displayed in the microscope. Here we have two things successfully accomplished; the striæ which in the former case were so " fine " as to be invisible have now become broad and can easily be distinguished by the eye; and, secondly,

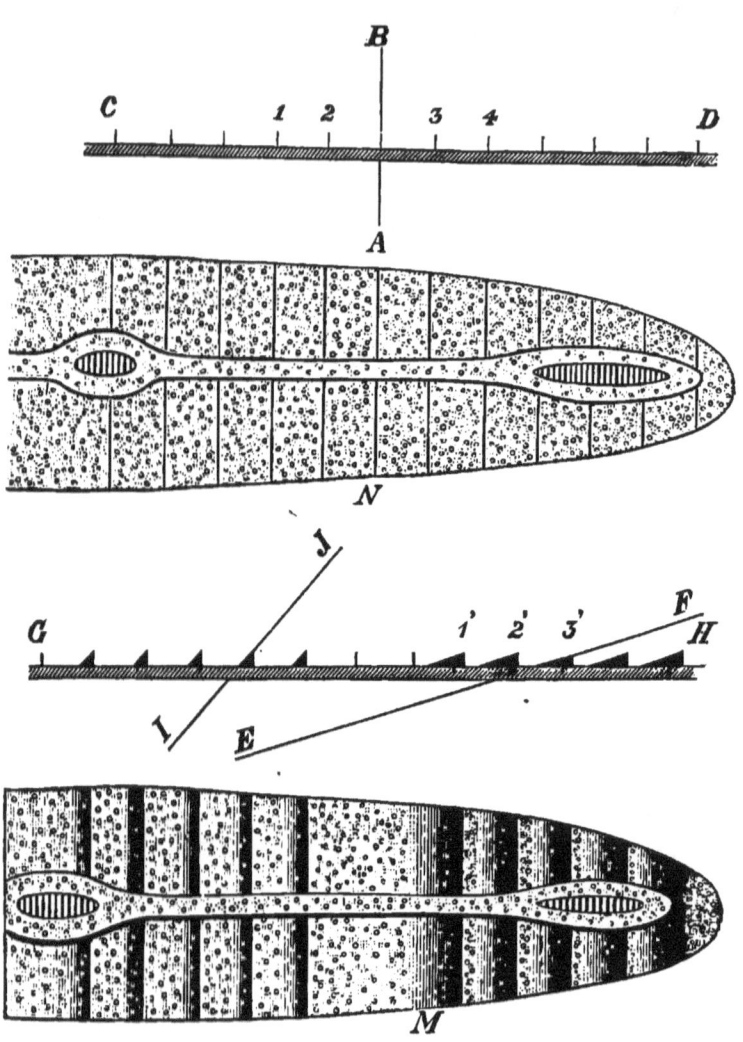

this last display is as *false*, as to many it has been acceptable. Now let down the obliquity of the illumination, as indicated by the line I-J; the effect is noted on the sketch at L, and with a first class high balsam angled objective and an eye well trained, it may be that the striæ are not only discernable in the microscope, but the observer will further be able to note the *intervening spaces* also; and this, too, which the certain knowledge that of the varying condition set forth, the latter display is not only the most satisfactory in general terms, but the most truthful. It is also apparent that the more we *decrease* the angle of the illuminating beam the louder the call on the defining power of the object glass and the greater the demand for education on the part of the eye. In other words, the expert, all other things being equal, with a pair of eyes trained by long practice in his profession, has the better chance of seeing things as they actually are. From the illustration given we deduce several propositions, viz:

First, it is always better to see structure *somehow* than not at all. Let those addicted solely to the use of narrow apertures ponder this well.

Secondly, when engaged in investigations of "difficult" structure similar to the case presented, and calling. at the best for light of considerable obliquity, the less of the obliquity, as a rule, and within certain limits, the better; and the more perfect the education of the eye the less will be the call above mentioned.

Third, the higher the balsam angle of the glass the LESS will be the obliquity required. Let those who favor work with centrally disposed light make a note of this.

Says one (alluding to the second proposition), I don't exactly see that thing as you do. Suppose, for instance, that the objective and the eye were both so perfect as to allow the illumination to become axial, as in your first illustration, wouldn't that be better still?

We reply, ask the artist if he would prefer the landscape thus lighted. The architect of the universe expressly arranged things to prevent such a catastrophe. A certain amount of shade is as necessary as that of light. Such a thing as " dead central illumination," although often talked about, is a myth.

Feeling deeply the importance of calling the attention of the reader to an element which we regard as of vital importance, we have accordingly done so at the risk of being somewhat tedious.

POSITION OF OBSERVER.

Our experience is, that in sixty cases out of the hundred, having made some preliminary examination of an object under study, and thus demonstrating the necessity of the use of higher amplifications, that with the latter comes also the necessity of a long and protracted sitting; in fact a downright seige is inaugurated. It is better in all such cases to postpone work until evening, or at least so to arrange that the observer shall have perfect immunity from interruptions of any nature; and at the commencement of such work it is of paramount importance that the operator adopt such a posi-

tion at the work table as will allow him to observe for
hours without serious fatigue. Such a table as has
already been recommended, with two front drawers,
will be just what is wanted. The novice is here reminded that in working with the higher powers, after
having placed the object in position under the objective,
the latter being correctly adjusted, there will be no
longer use for the coarse adjustment. Having, then,
our table, place the chair adjacent thereto, but in place
of putting its front edge parallel to the front of the
table, as is generally done, turn the chair to the right
until the front and left-hand sides form equal angles
with the front of the table, the angular point formed
by the meeting of the front and left-hand edges of the
chair being adjacent to the table. Now let the observer
seat himself, placing the shank of the left shoe on the
left-hand round of the chair. Pull out both drawers,
so that the edge of these, assisted by the edge of the
table, shall form a double rest for the forearm. Now
place the stand in such a position, that is, with reference to the front of the table, that the left hand finds
its way easily to the fine adjustment, while that of the
right grasps the mirror bodily, the tube being meanwhile adjusted to the standard length, and the whole
instrument properly inclined. Let the reader practice
these directions thoroughly until he shall be able to
thus sit at his instrument firmly wedged to it. It will be
noticed that, once in this position, either hand can grasp
the object slide for the purpose of making any necessary change, and without seriously disturbing the double

adjustment for the forearm, while the right hand, being thus so nicely supported, is enabled to manipulate the mirror with almost mathematical precision. At the first, as might easily be supposed, the sharp edges of the table and drawers will be a source of some little inconvenience. This can be remedied by placing a pair of napkins thereon; but the better way is to endure this slight annoyance for a little time, when the forearm will be found to have adapted itself to the situation. The position thus described we have represented as far as possible in the frontispiece, using such furniture as was at hand in the photographer's gallery

This pose can be varied at times by bringing the left knee in use so that it may support the left elbow. Thus we get three rests for the left arm, and sometimes we get the shank of the right foot on to the front round of the chair, spreading the knee open a bit and wedging it under the right hand draw. Other little changes are practicable and need not be detailed here.

Of all the manipulations connected with the use of the higher powers, the adjustment of the objective (supposing, of course, that it is a good one, one that will *respond* to the collar adjustment), is of paramount importance. For reasons previously stated, two elements are involved, namely, the behavior of the object glass and the education of the eye; and here the use of the diatommecæ is imperative. We have insisted on this for years. These little organisms are the most convenient, and then, again, any little deviation from the perfect correction of the objective is sure to " stick out "

and be detected.   This is not the case with other objects.
Show the tyro a scale of poduræ under a tolerably narrow angled one-tenth, and likewise display the same
with a similar object glass of the higher balsam angles.
There will be as much difference in the quality of the
two exhibits as there is (I was going to say) between
light and total darkness; and yet, nevertheless, the novice will be as well satisfied with the one as with the
other; but let the experiment be repeated, using in the
place of the podura a balsam mount of surriella gemma,
both glasses doing their best, as before, and the tyro is
no longer " at sea " as to his choice.   Moreover, at this
stage of his experience he will be fully prepared to blow
his trumpet in the support of one of the most absurd
and stupid errors that has ever been promulgated since
the time of Adam, to wit, " High angled glasses are only
fit for work over diatoms! !"

When we say that diatoms are the most convenient
objects over which to study the adjustment of the objective, *we mean it*; and thereto attaches greater force
than the casual reader may suppose.   If it be imagined
that these objects are " convenient " because their general proportions are about the thing—because they can
be purchased at slight expense, or, if preferred, prepared by the observer himself, or even be it granted
that the markings on the more difficult of these shells
will only surrender to a first-class objective in perfect
adjustment, we admit the facts, but the story is not
fully told.

The grand, the culminating convenience attending

the use of diatoms in the study of the objectives used
under high amplifications, is this: We are enabled to
display on one and the same mount, shells of the same
family and species, differing only in size, and we thus
are on the instant ready to study the work of the objective over each. Now if it so be, and *we* make it a
point that it *shall* so be that the smaller shell is in all
respects the more difficult of the two. Then it occurs
that the student, having mastered such smaller frustule
can examine at his leisure the larger ones, and with the
certainty that his objective is in at least approximate
adjustment; hence he is further prepared to note the
difference in the behavior of the object glass over the
different diatoms, and thus he arrives at items of the
utmost value; all this, too, without any change of the
mount.

Still other conveniences there are attending the use
of the diatom, their extreme thinness preventing the
shadow of one shell from interfering with the definition of another, thus getting rid of a complication
which would prove of serious detriment in the *early*
studies of the student, while by and by he can essay an
attack on the very problem named by merely selecting
such positions of the amount as contain the little
organisms huddled together. Thus learning their character and being thus forewarned, is fore-armed against
the time when he shall be brought in contact with
other slides presenting the same difficulty, but in a
more determined manner.

But let the reader note this fact. It is one thing to

look at diatoms, and quite another to study them with the especial object of becoming acquainted with the behavior of the objective, while it must be admitted that there is a fascination and charm *per se*, connected with diatom examinations under the microscope. It is equally true that the student can use them legitimately and for the purpose named, without establishing any claim to the functions of the diatomist.

Referring to the especial purposes we have been considering, the following list of objects will amply suffice, for the study of the one-sixth or one-tenth objectives, viz:

1. Navicular Rhomboides, Monmouth, Maine, Balsam Mount.
2. Navicular Rhomboides, Cherryfield, Balsam Mount.
3. Frustulia Saxonica, Leipsig, Germany. Balsam Mount.
4. Frustulia Saxonica, Isle of Shoals, U. S., Balsam Mount.
5. Amphipleura Pellucida, Bridge of Allan, Scotland, Balsam Mount.
6. Amphipleura Pellucida, Aberdeen, Scotland, Balsam Mount.
7. Surriella Gemma, Balsam Mount.

Of the above list, Nos. 1 and 2 can readily be obtained of the dealers. As to the others named, Prof. H. L. Smith, of Geneva, N. Y., has a large supply of the material, and has kindly supplied the author and his friends with excellent mounts and at figures much below the usual list prices.·

In addition to the diatoms, a *genuine* mount of English podura will be a capital thing to have on hand for occasional comparisons, but the student must on no account select the slide himself. Poor scales are simply good for nothing, while perfect ones are held by the opticians in the highest esteem, and are constantly used

by them in the final corrections of the objective. Of
all the scales of podura we have yet seen, a mount was
shown us last summer by Mr. Herbert Spencer, which
was perfection itself. As an index of the value Mr.
Spencer attached to this mount, I will add that he had
once offered one of his first class one-tenth objectives
in exchange for it !

Every microscopist needs a suitable stage micrometer
and very few have reliable ones. We gladly state that
Prof. W. A. Rogers, of Cambridge, Mass., has after
years of careful study succeeded in making rulings on
glass rivaling if not excelling those of Nobert himself.
Not long since Prof. Rogers ruled for us a plate, containing lines 100, 1,000, 2,000, 5,000, 10,000, 20,000,
40,000 and 80,000 to the inch, which was a marvel in
point of accuracy and delicacy. Subsequently Prof.
Rogers ruled a plate for us up to 120,000 lines to the
inch. This band the author has seen well and has
shown to his friends. We therefore recommend first,
that every student should be possessed of a micrometer;
and secondly, that the same be procured from Prof.
Rogers. It will be well while one is about it, to order
a plate ruled as high as 80,000 to the inch, inasmuch as
the cost is not materially enhanced, the plate will thus
do double service not only as a stage micrometer, but
as a test plate for the comparison and study of objectives. In this connection the reader will remember
that we have advised that eye-piece micrometers be
ruled 600 lines to the inch. Hence, if our advice is
followed, let the stage micrometer have a similar band.

The positive convenience resulting from this in the measurement of the focal length of objectives by the method previously described is too obvious to need further comment. The series of graduated diatoms by Moller of Wedel, Germany, and generally known as the Moller test Plate, is now to be found in the cabinet of nearly every microscopist, and can be advantageously supplemented to the list above given. A table showing the mean of the measurement of ten of these plates will be now furnished and thus the plate can on a pinch be made to do approximate duty as a stage micrometer.

MEAN OF TEN MEASUREMENTS OF MOLLER TEST PLATES,

BY PROF. E. W. MORELEY, M. D., OF HUDSON, OHIO.

1. Triceratium Favus.................................... 3.1 to 4.
2. Pinnularia Nobilis...............................11.7 to 14.
3. Navicula Lyra...................................14.5 to 18.
4. Navicula Lyra................................... ....23.0 to 30.5.
5. Pinnularia Interrupta,.........................25.5 to 29.5.
6. Stauronesis Phoenicenteron....................31. to 36.5.
7. Grammatophora Marina......................36. to 39.
8. Pleurosigma Balticum.........................32. to 37.
9. Pleurosigma Acuminatum.....................41. to 46.5.
10. Nitzschia Amphoyx...........................43. to 49.
11. Pleurosigma Angulatum.......................44. to 49.
12. Grammatophora Oceanica=G. Subtilissima ....60. to 67.
13. Surriella Gemma.............................43. to 54.
14. Nitzschia Sigmoidea............. ....61. to 64.
15. Pleurosigma Fasciola.........................55. to 58.
16. Surriella Gemma. Longitudinal...............64. to 69.
17. Cymatopleura Elliptica.......................55. to 81.
18. Navicula Crassinervis, Frustulia Saxonica.....78. to 87.
19. Nitzschia Curvula.............................83. to 90.
20. Amphipleura Pellucida.......................92. to 95.

THESE FIGURES DENOTING THE NUMBER OF LINES IN .001
OF AN ENGLISH INCH.

As a matter of convenience to those having the celebrated Nobert 19 band test plate, we present the following tabulated values of the rulings. The first
column contains the value in Paris lines. The second
the number of lines in .001 of an English inch, as ruled
in the several bands.

NUMBER.	PARIS LINES.	IN .100 ENG. IN.
1	1,000	11.26
2	1,500	16.89
3	2,000	22.52
4	2,500	28.13
5	3,000	33.78
6	3,500	39.41
7	4,000	45.04
8	4,500	50.67
9	5,000	56.30
10	5,500	61.93
11	6,000	67.56
12	6,500	73.19
13	7,000	78.82
14	7,500	84.45
15	8,000	90.08
16	8,500	95.71
17	9,000	101.34
18	9,500	106.97
19	10,000	112.60

Returning now to our original list of diatom mounts
we proceed to give such a description of the same as
will enable the student to approach them intelligently,
and without undue loss or waste of time. The first two
slides are pretty much the same thing, both contain

fine shells of navicular rhomboides, and nicely assorted
as to size and "difficulty." The larger, or even the
medium sized shells are easily displayed (we refer to
the transverse striæ) by such a glass as the Spencer dry
one-fourth of 115°, when nicely adjusted, while the
smaller frustules are more "difficult" than the No. 18
of the Moller plate. These mounts are therfore valuable on account of their range, and the reader has
already our reasons governing the selection thereof.
When these slides are thoroughly mastered, the student
will be ready to attack the Saxonica from Leipsig.
These vary also, a medium shell being about as obsti-
nate as the No. 18 of the Moller plate, and will defeat
the Spencer one-fourth. The Saxonica from the Isle
of Shoals is still more bothersome. These frustules
are very small and very thin, and at the start the
student should select the very largest and most vigorous
shell he can find on the mount, proceeding carefully as
he may acquire skill, to the smaller ones, and before he
can establish any claim to have mastered this slide he
must be able to show the very smallest and thinnest
shell acceptably. In point of "difficulty" these smaller
diatoms rival the No. 19 of the Moller plate, while the
smallest and the very thinnest are a fair match for the
No. 20. Of the amphipleuras, that from the "Bridge
of Allan," Scotland, is the easier of the two, and
should be studied first. This slide in common with the
two last mentioned calls for object-glasses of high
apertures. To attack them with any of the low angled
glasses is but a waste of time and patience. The mount

of surriella gemma is selected for reasons very much
akin to the mention that has been made of podura.
These shells are not suited for the purpose of study,
but are admirably adapted for comparisons. The
actual structure of this diatom is yet a matter *sub
judice.*

Now let us go back again to the first mounts named
in the list. We have had occasion before to assert that
one of the advantages resulting from the study of diatoms is due to the fact that the student in mastering
one of the smaller and more difficult ones becomes
fortified as to the adjustment of his glass. Supposing
then that by some good luck he has succeeded in displaying the smaller frustules of the Monmouth or
Cherryfield. He can return to the larger shells with
tolerable.assurance that his glass is in fair correction.
But mark this. In order to show the little fellows,
there was need of greater obliquity of illumination,
which, when return shall be made to the larger shells
can profitably be dispensed with. Now although the
slide of diatoms offer special conveniences, let the
student adopt this line of study, no matter what may
be the character of the object under his objective, selecting the most obscure and diaphanous structure possible to bring into the field, and making sure that there
is a sufficiency of oblique illumination.

From what has been said it may be adduced as a general rule that it is advisable to study the corrections of
the objective by the aid of oblique light, and over the
finest structure the mount can be made to exhibit, and

to the beginner no better advice can be given. It is simply impossible for the novice to study the corrections of a high angled object-glass by centrally disposed illumination.

In the selection of the objects recommended to the student, we have been governed by these and similar considerations. All of the slides named contain frustules requiring the use of light at least 40° from axis, and the student will in his first attempts be compelled to go considerably higher than that figure. The exact amount of obliquity essential to the best display *he* can possibly make, will depend on his object-glass the correctness of its adjustment, and the education of his eye.

We shall, on some succeeding page have a few words to say as to the management, etc., of high balsam apertures by centrally disposed light. With this one exception only, anything that we may have to offer in the way of instruction bearing on the use of these glasses will be confined to the study of the diatoms which have been presented. The student who can properly display the seven mounts to which his attention has been invited, together with the Moller plate and the podura, can safely be allowed to shift for himself. We, however, fear that our ability to teach this much will not now keep pace with our sincere desire. Competent, or incompetent, we propose to *try*. A singular fact it is, that among all the works thus far written on the microscope not one word of instruction can be found as to the management of objectives of wide apertures. It is therefore high time that some one "assume the judicial." *Any* endeavor on the part of *anyone* to teach the management of wide apertured object-glasses must in the nature of things be incomplete, and at the best but approximative. Notably so as to written instructions, no two objectives work exactly alike, and appearances in the field will be modified by the individual corrections of the particular glass employed. Of two objectives, one may be superior to the other. Hence the student not being a connoisseur becomes environed with doubts that the expert might easily cause to vanish by a little personal interference. Object-glasses, too, work differently over different objects. The appearance of a dry mounted object is characteristically at variance with the appearances presented by balsam mounts, and here is our reason for selecting in every instance the latter for the instruction of the beginner. Neither is it possible for the professional philologist by mere force of word picturing to convey to the mind of the learner just what may be desired. And even should the teacher be fortunate enough to employ the most accurate descriptions language can afford, then he has no assurance of being perfectly understood by the pupil. The author being no philologist, and having but a little command of language desirous of using none other than the simplest words, thus adapting this little book to the comprehension of all, can but essay his best and only hope for partial success.

A word or two further before we proceed with our

tuition. The remark is often of late made, "what does
all this talk about the handling " of objectives mean?
Are we to understand that the real work of the microscopist consists in the resolution of a few diatoms, or
the exhibition of the rulings of the 19th Nobert band?
To this we reply, that all this is implied and a very
great deal more; it means that the microscopist shall
know as much of the microscope as it is expected of
the engineer to know of his level or transit. In
geodesy the value of entire systems of costly triangu-
lations are wholly dependent on the accuracy of the
original base line, and in a similar manner is the work
of the microscopist affected. We often hear old observers claim that all work worth having with the
microscope is accomplished with amplifications less than
300 diameters; and then again the scientific contributions resulting from the use of the finest English objectives in the hands of physicists like Professors Tyndall
and Huxley dependent on amplifications of 5,000 and
6,000 diameters, have been seriously called in question,
yea, disputed by those working with objectives not
worth the weight of their brass mountings! We
repeat, that it *is* eminently the business of the would-
be microscopist to know all that can be known of his
instrument; if, when he has to an acceptable extent acquired this knowledge, if so be that he prefers not to
devote his time and microscope to investigation, he has
the satisfaction of being conscious that what was done
was well done. Had the engineers employed in measuring the original base on which the coast survey was

founded, chosen to have abandoned the work on the
completion of the base, there would have been no time
lost or wasted; *per contra*, had the base been but imperfectly measured, and the engineers proceeded to
build the other triangles resting thereon, they would
not long have continued in the service, but would have
been furnished with palpable reasons for leaving!

Again, the question is ofttimes asked, do you require
that the observer shall be familiar enough with his objective as to work it up to the same force as its maker
could? Its the legitimate business of the optician, and
he can thus afford to spend his time on a matter which
to him is a necessity, but to demand this much of the
practical observer seems uncalled for. In answer to
this oft-repeated remark, we wish it distinctly understood that it is eminently the business of the observer
to *fully* understand the practical working of his objectives. The man who cannot make his object-glass perform as well as when in the hands of its maker, is using
an instrument that he does not fully understand, of
which the fact becomes incontestible proof. If it be
necessary that the optician be able to work an objective to its maximum, it is imperative that the microscopist should do as much. Until this shall be the
case, absurdities will abound. Take for instance a condition that often occurs. The microscopist orders a
new objective from the optician. The specifications are
complete, and the glass is made conformably thereto.
In due time the purchaser receives his glass, and can do
nothing with it, and it is returned in high dungeon to

its maker. And why? Simply because the purchaser was as incompetent to "handle" the instrument, as perchance he might have been to manipulate a pianoforte or church organ. And then again through his ignorance he might have possibly ordered the very glass he did not want. What would we say to the invalid who on summoning the attendance of the physician should undertake to dictate as to the treatment!

Again, we have often been requested by our friends and visitors to examine objectives and give our opinion thereon; and, as a rule, five minutes will suffice for this. But, says the owner, "you must be prejudiced against my glass. No man could give a well grounded opinion in the few moments you have occupied." Nevertheless, we have given hundreds of just such opinions without having had occasion to modify or retract, and here is an illustration showing that time spent in the study of objectives is neither totally lost nor without its compensations. We reiterate, let the microscopist understand *well* his tools, and in the hope of being at least some assistance to the student we are ready to proceed with our instructions.

In the lessons which follow the reader is presumed to employ a wide apertured one-fifth, one-sixth or one-tenth immersion. We shall try, as far as possible, to make them applicable to the use of any good glass having air angle up to 175°. The higher balsam angles will, however, when attainable, be the better instrument, with which, too, the instructions will be the better understood. The student is also reminded that with the

particular objective he may employ he may be in some
measure defeated in his efforts; still we have to hope
that his endeavors will not result in time "totally
lost."

Lesson First.—Place the mount of Monmouth rhom-
boids on the stand, select the two-inch ocular, making immersion contact with water or glycerine, as has
previously been directed. Place the collar of the adjustment in the middle of its run; use the light from
the smallest kerosene hand lamp, placing the same about
seven inches to the left of the stage, the flame to be on
a level with the same, edgewise to the mirror, and two
inches in advance of the front edge thereof. Removing
all sub-stage appliances, swing the radial bar so that the
mirror shall give illumination, say from 70° axis; manipulate so as to light up the field, using about one-half
the sized flame your lamp will allow. Now focus, and
having taken a general look at the "tout ensemble,"
select a frustule somewhat below a medium one in size,
and bring to the centre of the field, the valve in a horizontal position. Now focus again carefully, adjusting
the mirror so as to get the very best view attainable.
Now study for a moment the general character of the
image. Notice particularly if there be any distortion,
or whether the two ends are more or less indistinct or
obscured. Slide the object carrier and hunt up several
other shells and examine individually, that you may be
assured that you have a fair specimen to deal with.
Next, notice the edges of the valve; compare the upper
with the under; note if these are tolerably clean and

sharp, or whether they are more or less woolley. Compare the median line with the upper and lower edges. Manipulate the fine adjustment and notice as to any change of focus necessary as between the edges and the median line; study this median line well, and you will probably detect a shadow from it shading, but only perhaps slightly, the lower entire half of the shell. Grasp the mirror firmly and change its position (*not* the radial arm) a little in every direction, meanwhile studying the effect of the shadow on the lower half of the frustule. Hunt up other frustules and study them in like manner, changing the focus from time to time as may be necessary.

Selecting an average shell, and placing the same in the centre of the field, and focus sharply. Having arranged the mirror to the best advantage, grasp the collar of the objective with the fingers of the right hand (which now leaves the mirror just as it was), those of the left hand being in contact with the fine adjustment, withdraw the eye for an instant and turn the collar of the adjustment briskly nearly a full half revolution; placing again the eye in position, focus quickly and sharply. Now undertake to decide instantly as to the effects produced.' Is there more or less distortion at the ends of the shell than before? How are the edges? Cleaner and sharper, or the reverse? How is that shadow below the median line? Does it shade the entire half of the valve more than was the case formerly, or has it consolidated itself to a narrow band under the median line? If you find the least trouble in assuring

yourself *instantly*, repeat the programme *until you can.*
Now, reader, one-half hour conscientiously spent at this
exercise will assist you more than a half year of pro-
miscuous practice. Let me advise you to hang *well* to
it. You must learn to judge *instantly* as to the appearances to which your attention has been called, nor can
the least " guess-work " be allowed.

Well, now, premising that you have had sufficient of
this practice, so that you are able to assert, without fear
of contradiction, that after the turn of the collar adjustment, there was, as we will assume, evidently more distortion noticeable at the ends of the valve, nor were
the edges as sharply defined, (it may be that one edge
cannot now be seen without specially focussing therefor.)
The shadow, too, under the median line is not only entirely over the entire lower half of the shell, but it has
become deeper and more offensive than before. The
fact, then, becomes apparent that you turned the collar
in the wrong direction. Therefore, place the adjustment in its original position; look your shell well over,
thus to fortify yourself. It will be well, now, to repeat
the experiment, selecting various valves of the rhomboides, and, having become accustomed to the change
in appearances, then you may try many of the other
diatoms to be found on the mount. No matter what
particular one may be selected, you will in due time be
able to note the characteristic effect of *this* change in
the adjustment.

Lesson Second.—Having replaced the collar adjustment to its first position, and selecting a tolerably small

shell of the rhomboides, place it in the centre of the
field, the focus well adjusted, etc. Devote a few moments to a general consideration of the nature of the
changes wrought by the turn of the collar. Endeavor
to gather up, as it were, the experience gained by lesson first. This done, removing the eye for a moment,
turn the adjustment in an opposite direction to that of
lesson first, but not so far in extent, and immediately
looking through the tube, adjust again the focus, and if
necessary adjust also the mirror until the best view of
the shell shall be obtained.

Now if the preceding lesson has been well studied, the
learner ought to recognize a decided improvement, and
there ought, withal, to be more general brilliancy to the
image. The shell should, so to speak, lay flatter in the
field. In short, things ought to assume an encouraging
appearance. Repeat the experiment several times, but
with greater or lesser changes of the collar adjustment,
until you arrive at a particular point in the adjustment
which seems to be about the thing. Having thus determined this point, the student may now, by way of en-
couragement, shift the position of the object carrier
and bring one of the larger shells to the centre of the
field, when, with a slight alteration of the mirror, he
will probably be rewarded with a fair view of the transverse striæ. Assuming this to be so, let the student
examine progressively other and smaller frustules, and
selecting one of the very smallest that he can see, bring
it to the centre of the field.

Lesson Third.—Replace the collar to the initial po-

sition, as per lesson second, focus your new shell and
examine it critically, as has already been directed; next,
repeat the turn of the collar, in the same direction, but
varying the extent of the movement, focussing again
as before. Bye-and-bye, when the collar shall have been
turned to approximately the correct place, the striæ will
of course be again visible, and it is more than probable
will be even better seen than before. If this shall be
the case, try, by slight manipulations of the mirror
alone, to obtain the very best display of the *lines*. And
we now approach close work. If the eyes are tired, or
the general condition of health be below par, it will be
better, if any further work be determined upon, to go
back and take a review of what has already been accomplished; but on the other hand, if conditions are generally favorable, proceed thus:

Lesson Fourth.—With your frustule as nicely exhibited as possible, the shell selected being as small as you
can successfully attack to show well the striæ, examine
it well closely; see if the upper and under edge are
both equally well in focus. It may be that one edge is
much better defined than the other, or it may be that
one edge cannot be seen at all. This indicates that the
valve lies sidewise in the balsam, in which event discard
the shell and hunt up another that lies exactly flat, and
of the same size.

Now work over this diatom, generally as you have
been instructed in the foregoing lessons, *accepting the
position you now have of the adjustment as the base of
farther trials*. Change the collar to the right or left.

noticing all the points to which your attention has been called; but these various changes of the collar should be much less limited in extent. We have now further items for you to observe. Examine the *striæ* carefully; see if they continue *quite* to the edge of the shell, or do they stop a little short thereof, the edge becoming as clear, sharp and distinct as that of a razor. Look with an eagle eye, too, *into* that shadow under the median line. Has it become almost self-luminous and transparent? And has it drawn itself upwards and together, forming a narrow but brilliant band (so to speak) adjacent to the under side of the median line? Have all appearances of diffraction lines left? Does the shell lay well down and flat? All of these interrogatories must receive your devoted attention.

When things are just right, the valve laying flat in the balsam, the striæ will not *quite* extend to the edge of the shell. The dark band under the median line will be contracted as described, but in the place of the former dark and muggy shadow, it will have become lively and brilliant. Ditto, as to all the other shadows seen in the field. In fact, everything has improved, be it a diatom or a patch of dirt. The edge of your shell ought to be sharply distinguishable, not by a line of varying thickness, for there is no line there, and therefore you ought not to see anything of the kind. If the adjustment be a trifle out, it will sometimes occur that the valve will seem to "rise up;" *i. e.*, appear nearer to the eye than it should. It is impossible to describe on paper just what I mean by this. I have no trouble,

however, in getting pupils to appreciate this phenome-
non. This effect is the more palpable under a high ocular, say the one-fourth inch. Let the student try the
highest one he has, and he will notice that a slight change
of the collar will cause the shell to " lay down " properly. But he must nevertheless be able to make this
correction from noticing the condition when employing
the two-inch eye-piece. And here it may be observed
that with different objectives some leeway must be
allowed. It is hardly probable that the student can
take his particular object glass and follow me to the
very letter; or is it even certain that the author could,
with said glass, demonstrate to an expert what has been
written in these instructions; yet after allowing due and
proper margin, the hints presented must prove of value,
and if we could have been told as much eight years ago
it would have saved us hours and hours of the toughest
work. And right here a thought presents itself, foreign, perhaps, to these lessons, but let it go on record,
namely: If it be so that we can assist the student, we
have in very truth no mean reward for time spent in
the past in the study of object glasses.

Lesson Fifth.—Having noted the division corresponding to the exact adjustment of your objective, so that
you shall be able to place the collar at the right spot at
once, without loss of time, leave the rhomboides and
examine your mount generally. Notice the play of light
and shade over the several diatoms. Hunt up those portions of the slide containing a mass of the larger forms
huddled together; in other places you will probably

find sediment congregated together in little colonies, the coarser particles throwing an individual shadow over the others. Observe these appearances closely. Now place the collar intentionally out of adjustment (preferably opening the systems), and thus look over your mount again. See, how the shadows have become mixed up! What a labyrinth there is to be sure! A perfect plexus of indeterminable, indefinable shadows proceeding from nowhere and ending nowhere. The thickened edges of the diatoms seem to cast a shadow on their own account. Notice, too, how that some of these large shells seem to have become badly distorted. Find those little colonies again, and note that the individual particles thereof have lost their brilliancy—have become muggy and indefinite. Here, too, are all sorts of shadows, and more or less confusion generally. Return, now the collar to its proper place and review your mount. Now, see how things have improved in general brilliancy; see how the shadows have become harnessed properly into the traces. You can trace them now, and get an idea of what they mean, where they came from, and where they end. Look at the little colonies once more. Now they are all alive with brilliancy and sparkle like very diamonds.

When the student feels that he can follow me thus far, let him study the *tone* of his *objective*. This will depend on the object-glass employed. All superfine objectives should be *under* corrected, *i. e.*, there should be a preponderance of the blue; but hardly any two good objectives are exactly alike, some being under-

corrected to a greater extent than others. Whatever
may be the correction of a given object-glass, there
will be a corresponding tone to the field, and it's the
business of the learner not only to know the fact but
to be competent to recognize the characteristic appearance presented not by the particular object viewed
alone, but by the entire field of view. And here again
comes some nice eye work. The artist can look at the
landscape and recognize the very "atmosphere"—not
so the shepherd's boy. The former might talk until
dooms-day to the latter as to " atmosphere."

Now every good object-glass has an "atmosphere"
of its own, peculiar to itself, and depending on the
individual corrections of the glass. A moment's contemplation of the situation teaches the difficulty attending any effort on the part of the author to make himself understood so as to be of service to the learner.
Nevertheless, the attempt shall be made.

Lesson Sixth.—Place the shell of rhomboides exactly
in the centre of the field, and adjust the object-glass.
Arrange the mirror with its radial bar, so as to illuminate at about 45° or 50° from axis. Focussing sharply,
examine well your diatom as to color; the chances are
that both blue and red are to be observed. Select in
turn several shells and thus examine, and finally in-
spect all the objects on the mount before mentioned.
A little patient attention will teach you that there is
something apparently due to the blue that does not
attach to the red. For instance, the shells may have a
slight lavender tint; as soon as one can detect any-

thing in this way that is presumed to be characteristic, then put the object-glass out of adjustment and focussing anew make the comparison. If the student has hit the proper effect, he will notice that with the glass out of adjustment, said effect has vanished. Should (as will most probably be the case with his first efforts) he fail to mark any characteristic difference in color or tint, replace the collar adjustment and repeat the experiment; selecting, as near as he can judge, some "probability," and again throwing the glass out of proper adjustment examine again. The process must be repeated time after time until the end shall be gained. When by dint of practice the learner begins to feel that he "sees the point" let him in a similar manner examine all of the other diatoms on the mount until he shall have become perfectly familiar with the *tone* of his objective.

Lesson Seventh.—Having mastered tolerably well the previous instructions, let the learner now turn his attention to the *tone* of his field. This is a finer "point" than that of the preceding lesson, and demands in turn more of the eye-training. The problem is in all respects similar to that concerning the tone of the objective, differing in this one element only. The student must now endeavor to recognize a particular tone to the whole field when the object glass is in perfect adjustment. The instructions are quite of the character as those of the last lesson, and need not be repeated. The two lessons might have been reduced to one, were it not that a little practice on the former

will, and for the reason named, be advantageous before
studying the tone of the field. These last two lessons
should not only be well studied, but the student ought
constantly to endeavor to improve in his recognition of
tone. Once having become tolerably, I may say, comfortably expert, its practice becomes a second nature
and will be persisted in from choice, meanwhile the
observer becoming daily more and more proficient.

This characteristic tone of the field is more apparent
in objectives of the highest balsam angles, with the
super-excellent object-glasses of the Messrs. Spencer or
Mr. Tolles, when the glass is exactly in correction, the
tone of the field becomes a peculiar and *exceedingly*
delicate shade of apple green, which one soon learns to
recognize by aid of the teacher.

In the studies we have thus far presented, and which
ought to enlist the occasional attention of the pupil for
at least a month or six weeks, it is taken for granted
that he will meanwhile use the instrument as he may
elect, and either for pleasure or such profit as he can
arrive at on his own account. There will undoubtedly
be other objects brought into requisition not named in
our lists. Now, whatever practice he may have of this
description, there are some general conditions concerning high-angled objectives which ought not to be un-
heeded, for example, let the learner discover as near as
possible the point of the maximum aperture of his
glass ; it will be well, too, that he pay attention as to
thickness of the covers used. These two points have a
direct bearing on the pupil's progress and must not be
neglected.

Hand the expert a strange mount of the Monmouth or Cherryfield rhomboides. It is possible for him without looking at it, and by the sense of feeling alone to state thus: " This slide cannot be used with the one-tenth and glycerine contact. Use water and with the collar nearly at closed. If the one-sixth be employed use glycerine, and the objective will correct within four or five divisions from the closed point." Either statement being found on the actual test or trial to be the fact. In truth, any such statement from the expert founded on his sense of feeling alone would be worth more than two hours spent over the tube by the novice. Hence the importance of the learner's acquiring the same tact ("*knack*" is the popular word), nor is there anything so very difficult in its acquisition.

Lesson Eighth.—Returning to the slide of Monmouth Rhomboides, and choosing a frustule somewhat larger than a medium one, place it in the centre of the field, adjusting the objective, and displaying the striæ, employing an obliquity of say 70° from axis. Next rotate the stage through an arc of 45°; if necessary centre the object again. It will now have assumed a position intermediate between the horizontal and the vertical. Attempt, by slight changes in the position of the mirror, to display simultaneously *both* sets of lines. The inch ocular may now be substituted in place of the two inch. To do this nicely, getting both sets with equal force, requires indeed some little "knack." The collar adjustment being already correct, the manipulation will entirely be confined to the illumination. Possibly it

18 Microscopy.

may be well to shove the lamp a little farther away from you; *i. e.*, a little more in advance of the front edge of the stage, or perhaps raising or lowering the wick, thus increasing or diminishing the amount of light, may be of service. This is all that can be written, and the learner must simply make the most of and help himself. Rest assured that both sets of lines are there, and if it be that they are not shown, the diatom is not to blame. In case of total failure the observer may select one of the larger shells, proceeding to the smaller ones carefully and by degrees. With the illumination at present employed he cannot expect thus to show both sets on frustules smaller than a medium one. After having mastered, as well as may be, the Monmouth, substitute the Cherryfield, readjusting the objective, as a matter of course.

There now remain, for future consideration, on the slides of Monmouth and the Cherryfield, the smallest frustules of the rhomboides. These we will pass over for the present, and in the next lesson proceed to attack the Saxonicas, from Leipsic, Germany.

Lesson Ninth.—The student will recognize the frustules on this slide by their similarity to the two slides preceding. The valves are on an average smaller and thinner, and in consequence weaker objects to deal with; as the saying is, they are more " difficult." Select one of the largest and most vigorous shells; bring to the centre of the field, and, employing the inch ocular, focus, the illumination being about 70° from axis. Now examine closely your diatom, and see if you can bring your past experience to bear. Without attempting to observe as to the striæ, reckon up the general appearances and form some verdict as to whether the glass is in adjustment. If you decide that it is not, make up your mind to what extent. This done, proceed, if necessary, to adjust, following the directions already given, bearing now in mind that all the phenomena previously described will be much fainter and less decided than was the case of the Cherryfield or the Monmouth.

Let the student assure himself that the shell before him lays perfectly flat in the balsam. If this is not the case, it will be impossible to see both the upper and under edges simultaneously. These frustules are extremely thin, and when one edge is the lower of the two in the mounting medium, it will appear to vanish out of sight. Accepting that the diatom is a favorable one for study, the learner will notice that the upper and lower edges, as the adjustment approaches the proper position, behave better than was the case with the Cherryfield or the Monmouth. This is owing, of course, to the diaphanous character of the Leipsic. He will therefore endeavor to render these edges considerably sharper and cleaner than could possibly have been the case when working over the former slides. Let the pupil observe closely the behavior of the shadow under the median line. As has already been remarked, the effects are quite similar, but in a less marked degree. The eye must be educated to the situation. Notice, too, that as the adjustment becomes more perfect this shadow contracts, assuming the form of a narrow band immediately adjacent to the median line, while the entire lower half of the shell is of a lower *tone* than that of the upper.

Should the tryo have got along thus far tolerably well, it is more than probable that he will have been rewarded by at least a glimpse of the transverse striæ. He may now, by minute changes of the mirror, and by raising or lowering the wick of the lamp, endeavor to show the markings as strongly (that's a bad word) as neatly as possible. Having thus succeeded, look sharply and see if the lines extend quite to the edges of the frustule; it may be that some of them will be found to project; in either case something is wrong. Look first to the illumination, and, observing the shadow line under the median line, endeavor to contract and render this as narrow as possible. If this does not prevent the striæ from showing to the extreme edges of this valve, then recourse must be had to slight alterations in the position of the collar adjustment. It may be advisable to try a stronger ocular, say the " D " solid, if one is at hand. If under this the lines seem to " rise," try and correct this, as before instructed.

This apparent " rising " will perhaps strike the reader as something novel. Certain it is that we have never seen anything of the kind in print. The effect was, however, noticed long ago by the author, and it soon became one of his " points " in the tuition of his pupils, the latter, at the onset, seldom ever get his meaning. Nevertheless, in a very little time they " see it " plain enough. We have not command of a word that will precisely and accurately express what we desire. It

may be advantageous to the student if we repeat ejacu-
lations such as are heard from scholars, thus: "I can't
make this shell lay down." "I have a tolerably good
show, but the valve is restless." "You seem to anchor
your diatoms; mine are all out at sea." "The valve
will float in spite of me." "I have it; the shell lays
as quiet as a summer morn." "This valve seems to
come right up to the eye-piece," (referring generally to
the one-half or one-fourth inch.) Such and similar expressions we have heard over and over again. The
student can select the one that will be of the most
service.

The tone of the objective should be studied with the
Leipsig, just as has before been noticed when discussing
the Monmouth and the Cherryfield. The observer may,
at the commencement, entertain the notion that in regard to tone, the objective behaves very differently when
over the more difficult mount. If, however, the preceding lessons have become perfectly familiar there will
be but little trouble in recognizing the same peculiar
and characteristic tone not only pervading the valve,
but the entire field of the instrument.

Suppose, now, that the manipulator, having followed
the author with satisfactory success through the Monmouth and the Cherryfield, should fail thus to do when
the Leipsig is taken in hand, thus: There appears a certain indistinctness accompanying the images of the latter,
besides, the valves seem considerably distorted, and the
band of light behaves badly; so much so that the directions cannot be followed with any degree of certainty,

and that several trials have been ineffectually made; all of which lead to the same result. In this case the probabilities are that the objective is at fault, and that any attempt to work it by anyone, however expert, will only result in a waste of time. The student ought not to decide definitely until by practice he shall be entirely competent to judge for himself. We mention the fact here, because we know full well that some who may read these pages will try and follow the author, employing such objectives as they may have at hand. Now it might occur that an object glass capable of showing a medium Cherryfield or a Monmouth very well indeed, even cutting the larger shells into checks or squares, responding, too, to all the conditions we have presented, and yet entirely fail when worked over any but an exceptionally large shell of the Leipsig.

It may be *apropos* here to discuss another point which is suggested as we write: It will be noticed that all along in the course of these lessons we have made it obligatory that the simplest illumination shall alone be employed, reserving for the present any allusion to special methods used in the resolutions of the severest tests. Our reasons are patent. First, the little lamp recommended is in ordinary use, and is always at hand when wanted; as to its size, the small lamp is even superior to a larger one, while the smaller model has the advantage of burning less oil, is handier to manipulate, and is the more portable of the two. In a former part of the book we have acknowledged the superior force due to modified sunlight; nevertheless, the latter is unfit for the everyday work of the microscopist. The observer who depends thereon leans on a treacherous staff. Could we command sunlight, this objection would certainly have less force; but even in that case we would be compelled to fall back on artificial illumination in the evening, and eight-tenths of all microscope work is done after sundown. Hence it is of profound importance that the microscopist select such illumination as he can at all times control, and to learn to make the best possible use of it. Again, all that we have said as to sunlight applies with equal force to objectives that *require* to be worked with its aid. Let the reader *note well this fact.* It has been our purpose here, as it is every day with our pupils, to teach the use of the simplest methods, both as regards illumination and other items of management.

Returning, now, to our diatom slides again, we remark that the objective failing, with the stipulated illumination of showing the Leipsig, might be forced to do so tolerably well by the assistance of monochromatic sunlight. Admitting this, we still assert that such a glass is unfit for the purposes of the working microscopist beyond those it will respond to under the conditions which we have presented. It further obtains that the reader who, with such a glass, has followed us through the Cherryfield and the Monmouth, can proceed no farther until we are ready to change the illumination. Presuming, then, that the objective is quite competent, and that the observer has thus far followed us in a satisfactory manner, we attempt some further hints and suggestions.

Lesson Tenth.—Having succeeded in displaying the
striæ of the Leipsig, selecting one of the larger and
most vigorous valves, it will be instructive to notice
once more, and particularly the edges of the shell, if
the frustule is flat and the objective equal to the work.
There ought not, as has been before intimated, be the
slightest indication of any *line* bounding the shell, such
as would be seen in a drawing on paper; but in place
thereof (so to speak), on looking closely, there may be
discovered, when things are *just right*, an exceedingly
fine band of peculiar light, the exact tint of which will
vary with the particular objective employed; and, as a
rule, the most brilliant tone will be noticed on the upper
half of the shell.   If necessary, manipulate the mirror a
bit, and endeavor to " gather up " this band so as to
restrict it to the narrow space bounded by the very
edges and the termination of the striæ.   This band or
tint cannot be well seen on the Cherryfield or the Monmouth, unless, indeed, on the very smallest shells, which
for the present have been held in reserve.   Should there
be any tendency of the " lines " to " rise up," try a
deeper eye-piece, and, shifting the adjustment a trifle,
seek to remove the trouble, returning to the inch ocular
as quickly as practicable.

This pale boundary tint may, with different glasses,
assume quite different tints.   With one it may appear
as a pale blue; another will show it of a delicate fawn
color, or even a pale pink.   Objectives having the most
perfect corrections (I do not mean the most achromatic)
will give this band in what I call an " apple green,"

and the tone of the field will be the same as that of the band. The student should practice until he shall be competent to recognize the band and its accompanying tint according to the particular glass employed. This done, let him throw the glass a little out of adjustment, and it will be seen that the band has disappeared, and with it so has all the nice definition pertaining to the edge of the diatom. Readjust the glass and watch for its return as you manipulate. There will be no time wasted at this.

A remark of a general nature is in place here. In working over shells as coarse as the Cherryfield or the rhomboides, if the glass be well adjusted for one valve it may be accepted that it is sufficiently so for all on the mount. This is neither strictly nor practically true. As some of the diatoms lie nearer the cover than others, the adjustment requires to be changed in accordance with the thickness of the intervening strata of balsam; and on fine work, like the medium frustules of the Leipsig, this point must be kept well in mind.

Now, *with this point in mind*, let the student, returning to the Leipsig, adjusting his glass over the same shell we have had in mind, getting things just right " to a dot," let him slide the object carrier and hunt up another and similar valve; that is to say, one that lays just as flat in the balsam, and of the same size. Now focus, and endeavor to say positively whether the new shell is the same depth in the balsam as was the former. If it shall be the same depth, that edge band will be seen; otherwise not. Practice this *well*, hunting from time to time all the larger frustules to be found on the mount.

Before proceeding farther we have to say that the foregoing ten lessons must be well studied and thoroughly understood. They have been arranged and presented so as to lead the student safely and progressively through what has proved to many a labyrinth. We have hinted at the difficulties which beset us and saturate (so to speak) any attempt of ours to render our instructions perfectly intelligible to all. It follows, then, perforce, that defects and shortcomings in the way of elucidation ought to be compensated for on the part of the pupil by his vigorous determination to master the situation. Again, in the exercise thus far given the intention has been steadily held in mind; first, to present as plain a chart for the guidance of the learner as was within our power; and, secondly, to keep well in hand the necessary education of his eyes; and in the latter acquisition the element of *time* is quite as much a factor as that of diligence and determination. We strongly advise, then, the learner to confine himself to the suggestions already given, until he shall feel himself thoroughly and practically familiar with them.

Lesson Eleventh.—We will now study the Monmouth or the Cherryfield (either will answer our purpose) once more. Place the same on the stage, illuminate at an angle of, say seventy degrees from axis. Bring one of the largest shells to the centre of the field; adjust the glass and focus, getting the very nicest display of the transverse striæ obtainable. Now gradually let down

on the angle of the illumination little by little, adjusting the mirror so as not to lose the show of the lines.
Now if the student has studied *well* what has thus far
been taught, he ought to be astounded at two conditions
which will flash on his intelligence. In the first place,
he will find that he can decrease the angle of the illumination down to, say 35°, all the while holding the striæ
well in hand. The probability is that he has now too
much light, and it will be well to try the effect of turning down the wick of the little lamp a little. Having
found the very lowest angle of the illumination that
will command a full show of the striæ, return the mirror just a trifle—just enough to reinforce the display of
the lines. Compare the effect with that of the previous
lessons. Notice that your striæ shell and all have become more transparent. There is not even a suspicion
of oblique illumination. Your diatom (and all the
others decently in focus in the field) swim in a sea of
fire. That's a cold word—"lightning" expresses the
idea better. Observe how flat the shell is! Not one
particle of distortion. Observe, too, that shadow we
have before called attention to, under the median line.
See how transparent it has become. Did you say that
you cannot see it at all? Yes, it's there; look sharp
for it! Now look down *between* the striæ. See how
nicely, how *exquisitely* they are cut apart. The spaces
between them are as visible as the striæ themselves.
Notice, too, how that the most infinitesimal flickerings
of the lamp flame plays with consummate grace in and
among the lines. Look well to the edges of the valve.

Watch, when the flame flickers, for that little band of light, and when you catch it remark the wondrous delicacy of its tone. Note, too, how the flickering of the flame bathes the field with minute differences in tone, things are almost alive. Compare what you now see with anything that you saw a month ago. Has not your time been well expended?

When the student has arrived at a position that will enable him to fully appreciate what I am now writing, he is safe. No matter should he have been accustomed to the use of low apertures from childhood, he can never be forced to return to his first love, nor will he have trouble in following me to the very end.

Should the learner unfortunately find that in this lesson he cannot see things as described, and presuming that the objective is competent for the work, then it must occur that there is either trouble with the manipulations, or his eye is not sufficiently trained for the work in hand. Hence he cannot get the mirror *far enough* downwards; *i. e.*, he is compelled, in order to see the lines at all, to use light of too much obliquity. There is but one remedy (if he is sure as to the capacity of his objective, which he ought, perforce of the preceding lesson to demonstrate), to wit: Practice until the end shall be arrived at.

This effect, due to the lesser illumination, has, as the reader will remember, received our attention on a foregoing page. It may be advisable that he review what has been said, and thus combine theory with practice.

Lesson Twelfth.—We are now to deal with the more

difficult class of objects, such as are recognized as the
severest tests. It will be much better for the student
if he confine himself to the instructions already given,
until he shall have become perfectly at home with his
objectives and mounts before paying attention to what
we shall here have to offer. In the remainder of the
lessons the education of the eye will be an indispensable
condition to success, and the instructions can only be
understood acceptably by those who have paid rigid
attention to the preceding instructions.

Arrange the stand with the inch ocular, lamp and
mirror, just as described in lesson first. Get out the
large bull's eye condenser and interpose the same between the lamp and mirror, the flat side towards the
lamp, and within a couple of inches from its chimney,
adjust the height of the condenser so as to throw the
light on to the mirror. The beam from the lamp may
be slightly inclined upwards, but never downwards. A
horizontal beam is not objectionable. Now experiment
a little. Try and get the lamp as closely to the left of
the stage as practicable, leaving room to work the condenser. Next, place the slide of Cherryfield or Monmouth on the stage. Now reflect for a moment. Where
is the point of maximum aperture of your objective?
And how will the cover of your mount respond to it?
If its point of maximum aperture be at nearly "closed,"
and your cover of such a thickness as will allow the
glass to "correct" nearly at closed with glycerine, then
you are to use the same. If so be, however, that the
cover is thick enough thus to cause the glass to correct

near closed with water as the intermedium, then, of course, use water; or it may occur that the cover is an *exceedingly* thin one—so thin that even with the glycerine the correction would obtain at some considerable distance from closed, in which case select another very thin cover to supplement that on the mount. All this has been discussed before, but in the preceding lessons one might have got along tolerably well without particular observance of these conditions, but not so with the problems we are now about to tackle.

Having then decided the above point make the immersion contact, and selecting a medium shell of the rhomboides, bringing it to the center of the field focus. Arrange the mirror so as to get the strongest illumination; let the lamp burn with a moderate flame; if now there is obviously too much light, a perfect glare, bring the lamp closer to you (keeping the flame edgewise to the mirror constantly) moving the same in a line parallel to the left hand edge of the stage, readjust the condenser and mirror thus lighting up the field again. It will be seen that the farther the lamp shall be thus moved the less will be the illumination. The point to stop at is when you have shut out the superfluous sheen or glare. All this time we have supposed that the flat face of the condenser was parallel to the face of the mirror. Now change the position of the condenser causing it to assume a slightly diagonal pose, the nearer edge to be swung *away* from the stage; when the condenser is exactly right the mirror will have the appearance of being traversed by a *line* of light, that is, it will *not* appear to be equally illuminated as was the case before. Now look to the adjustment of the objective, and by slight manipulations of the mirror display the striæ, which will be right straight along work. Seize the base of the condenser, stand firmly, supporting the left forearm on the edges of the draw and table; move the condenser just a bit to and from the edge of the table, noticing the effect in the field, getting it thus in the best position possible. Look to your mirror again and see if that " line " is still there; this cannot be dispensed with. Now if the directions shall be strictly carried out, the display will be much finer than possibly could have been obtained with the former illumination, the diatom will seem to swim in a lake of fire, or as my pupils sometimes have said, " chain-lightning."

Do not be satisfied with any apparent success, but repeat the method of illumination, time and time again, comparing notes. The student will not probably meet with perfect success the first evening, although it may be that he will be quite satisfied with his efforts. If it so be that the condenser stands at quite an angle, more so than you think ought to be the case, never mind; if you have the right effects in the field, you can study the rationale at your leisure. The next thing in order is to study all the appearances which have been pointed out in the preceding lessons.

Now shifting the object-carrier, hunt up a valve next the very smallest on the mount, one that lies flat. If you are using a one-tenth your inch ocular will do, but

if but a sixth, then you will need the " D " solid, the
one-half inch. Try now and display the striæ on the
little shell; remember that they are at best but very
thin and faint. Nevertheless, if there has been sufficient practice with the " lesser illumination " previously
mentioned, there ought to be now but little trouble in
recognizing the lines. When once a good view has
been obtained, the observer will be astonished by a
little experimenting, to find how far the lamp wick
can be turned down—just allowing the flame to peep
out of the cone—and still retain nice views of the
striæ.

This entire position, including that of observer,
stand, mirror, and condenser, we have tried to illustrate in the cut on page 289, the diagonal pose of the
condenser we have intentionally exaggerated. It is
probable, too, that the student may find it necessary to
make some slight changes to suit the tools he may have
in hand. Let him keep this fact prominent, to wit,
although the condenser will give more light than the
illumination first recommended, nevertheless there
must be no more light used than is absolutely required
to see the object without difficulty as respects illumination. If it be desirable at any given time to employ a
higher ocular, always turn up the wick of the lamp; a
slight turn will be found quite sufficient, and while on
this subject let me say that nine out of ten microscopists use far too much light. They seem to be impressed
with the idea that unless the object is bedazzled in a
flood of light that it is not well shown. All this is in

POSITION OF OBSERVER.

error; when such illumination is employed the chances
are that all fine detail of structure will be drowned
out, while the effect on the eyes is most injurious. Keep
then the field just as cool as shall be consistent with
vivacity and *life*, show your objects brilliantly, and see
that even the shadows are *lively* and *transparent*.
Grasp the mirror firmly; get in the habit of this;
when you touch it at all *take hold of it boldly*. There
is considerable for the student to learn in the proper
handling of the mirror, so as to combine boldness with
delicacy and efficiency. Now as you change its position just a trifle, notice the action of the light. Does
it slide by as it were, without taking hold? Or is there
one particular position in which the light seems to
catch and " nip," bringing out the makings with a vim.
Now if the former is the case, there is something wrong
(the objective being in adjustment) and perchance the
whole system of illumination will require attention in
detail.

When the smallest shell of the Monmouth and
the Cherryfield have been thus mastered (as to their
transverse striæ) those of the Leipsig may be taken in
hand, and subsequently the smallest valves of the Isle
of Shoals; these are quite as severe tests as an ordinary balsamed amphipleura pellucida. The slide of
Leipsig ought to be *thoroughly mastered* previous to
attacking the Isle of Shoals; all the peculiar appearances which we have from time to time set forth, are
to be studied, and as the eyes become educated the
amount of obliquity of the illumination is to be de-

creased. This latter item is to be a constant study; bear in mind this one broad rule, that to use a wide-angled glass *well* is to work with the least obliquity possible, and when over easy tests, with *more central illumination* that can be effectively obtained with narrow apertured objectives. In short the employment of the wide apertures points to the use of centrally disposed light; this may be a novel doctrine to many, but not at all new to the author.

Strive then to work with the most central illumination possible; get rid of all shadows not indispensable; as a rule where the markings are dependent on thickness of structure, the greater this difference of thickness, *i. e.*, the more prominent the markings, the less will be the obliquity required; conversely the thinner and the fainter the markings the more obliquity is called for. Now the smallest shells of the Isle of Shoals are *extremely* thin, the striæ are not so very fine, probably not measuring closer than 85 in .001 English inch, but they are so thin and so fine as to beat out any object-glasses, excepting those of the higher apertures. And even with these the mirror will have to stand (artificial illumination being used as directed) at least 60° to 65° from axis; a fine one-tenth ought to show any and all of them, either with the one inch or one-fourth inch ocular. A similar one-sixth will require the half-inch.

Now, in the study of the last four or five lessons, the student is to constantly endeavor to improve himself in the adjustment of his objective. The broad land-marks have been laid down for his guidance, and these should not be lost sight of. As his eyes become trained to this kind of work, he will begin to pick up items of value on his own account. Especially will he arrive nearer a due appreciation of what we mean by tone of objective and field. He will thus be able to recognize for himself, confidently, too, when things are just right. Ditto, when they are *not* just right. Should the collar of his objective need to be changed, he must be able to make such change at once, and in the right direction. There must be no indecision—no guess work. He will have learned, also, that when the adjustment is exactly correct, not only has the display of the striæ become more satisfactory, but that the general appearances of the entire shells have improved. This effort at improvement of the adjustment of the objective should become a constant and never ending study; in truth, after one becomes tolerably advanced therein, it is no longer a " study," but rather a pastime.

We desire to insist with all the vigor we can command, that there is force attached to the reciprocal relations above named, to wit: That when the student, perforce of intelligent practice, shall be in position to assert, dogmatically, that his manipulations are correct, he will, conversely, be just as competent to reaffirm the fact when things are not in proper adjustment. When this obtains to an acceptable extent, he will be competent to use his glass for any field of investigation that he may elect, and, as a matter of course, he will regard those whom he is assured have not given proper attention to this essential item, knowing, too, that it is essential—"*cum grano salis.*" With our private pupils we have often tried amusing experiments, simply to become assured of their proficiency in this matter, thus: We sometimes place a drop of mucilage on the interior surface of the field lens of the eye-piece, replacing the same and watch for results. If the pupil is well trained there is no deceiving him. He will assert roundly that something is the matter. If I say to him, mildly, " why, that's pretty well, isn't it?" He replies at once, in tones that there is no dodging, " No, sir; there is something wrong, sure." And then, again, I have had the pupil pay me back in my own coin—placing a similar drop of mucilage in the interior surface of my own oculars.

We have already said that no two lenses work exactly alike. It remains, therefore, for the pupil to make a specialty of his own object glasses, first assuring himself, preferably by the advice of some expert friend, that he shall not waste time over an inferior glass. While he thus becomes more and more familiar with his own objectives, he will also acquire a general knowledge of all, and will in due time be able to take a strange object glass and work it nearly or quite up to its maximum.

Unless the observer be provided with a really superfine objective, he can hardly make much headway in the examination of the smaller shells of the Saxonica from the Isle of Shoals. Should he have much trouble in getting the striæ on the very largest frustules, these appearing quite obscure and of a generally uninviting aspect, the edges badly defined, and more or less distortion prevailing, he may be pretty sure that his glass is at fault. His time will therefore be much better spent in the examination of such objects as come within its powers. This point is well worth close attention, thus, perhaps, saving a double outlay of time and patience; and there has been a vast amount of valuable time wasted in fruitless attempts to resolve severe tests, for which purpose the objective was totally unfit and incapable.

The illumination we have just described is quite sufficient for the display of the transverse striæ of either slide of Amphipleura. As we have before stated, the "Bridge of Allan" is the easier of the two and should be studied first, and the student should give preference to the cleanest shells on the mount. If the smaller scales of the Isle of Shoals are well seen, and the "Bridge of Allan" resist, the fault will certainly be in the adjustment of the objective over the latter mount; possibly it may occur that the observer has selected an exceptionally difficult shell of the Pellucida. The better way will be to look over the slide in detail, of course, as before advised, giving preference to the cleaner and most inviting frustule.

If so be that the student has the Moller probbe plate, the illumination given is also quite sufficient to display the Nos. 18, 19, and 20. After he shall have become well advanced, there are some particular advantages connected with the study of this plate. He ought to see the Nos. 18 and 19 with but little trouble, and in a little time with the radial arm at 45° from axis. On this

plate passing from No. 19 to 20 is quite a jump. One
may see the No. 19 in a very satisfactory manner, and
yet fail of the No. 20. At the onset it will be better
to secure all the obliquity of light possible. Should
the Acme or the little Histological stand be used, the
upper one-third of the mirror may rise above the stage,
the lamp and condenser to be adjusted with the greatest
nicety. If the No. 20 resist the attack, it is preferable
not to spend much time over it, but go at once back to
the No. 19 and endeavor, by some careful re-adjustment,
to get a still better view of this; and when satisfied that
such has been accomplished then return to the attack
on the No. 20. One will always succeed much faster by
trying to improve what is seen than by blindly working
over an object of which the desired details cannot be
displayed at all. The ease with which the learner can
run from one test to another renders the Moller plate
very acceptable to the learner. Should it occur that
the No. 20 make a protracted resistance, let the above
programme be adhered to, spending nearly all of the
time in perfecting the display of the No. 19. It may
be worth while to add that the Moller Amphipleuras are
about on a par, in point of difficulty, with those from
Aberdeen, Scotland. If there be any difference, my
impression is that of the two the Moller No. 20 are
perhaps the easier test.

In studying the Aberdeen, or in fact any mount containing severe tests, it is always advisable to correct the
glass approximately by selecting some fine object, of
which the details can be displayed by such approximate

adjustment. For instance, on the "Aberdeen" it is probable that ome specimens of Nitzschia may be found. by hunting for them, and this should be done at once, and the glass corrected as nearly as possible. An approximate adjustment will suffice to reveal the transverse striæ, and the initial display ultimately improved by force of a little well directed manipulation. It may occur, too, that some of the nitzschia are more difficult than others. This being the case, the student is to master these successively before attacking the amphipleuras.

In the practice of all the preceding lessons, and over the test slides, it will be essential that the pupil, after having adjusted his glass as perfectly as may be, note the exact division of the collar graduations. This he should make a memorandum of, so that he can be able, if necessary, to readjust the objective to the mount without loss of time. But this is not all. He will find that as he acquires proficiency by practice, that he will, time after time, change these recorded numbers, and will be inclined, perhaps, to smile at the wildness attending his initial attempts; and by thus comparing the present with the past, he may be encouraged as to the future. This habit of noting in black and white the best attained adjustment of the objective from day to day should be rigidly persisted in for at least a twelve-month.

WORK OVER DRY MOUNTS WITH HIGH APERTURES OBJECT-IVES.

After, and not before, the student has become profi-cient in the preceding lessons, it will be advantageous for him to procure dry mounts of the acknowledged diatom tests, and study well their peculiar action under the objective. At a glance it will be noticed that the dry frustules are much more vigorous than those mounted in balsam. They have more body, appear more solid, the image is stronger every way. Hence it is that an objective may shew a dry mount tolerably well, and yet be utterly defeated by the same valves when balsam mounted. Should the student be well versed in our pre-vious instructions, he will find little difficulty in dealing with dry mounts. As a rule, the latter are to be illu-minated with pencils of less obliquity than objects mounted in balsam. In the dry mount, too, we have greater contrasts of light and shade. These differences will be at once noticed by the intelligent pupil, who will find little difficulty in adapting himself to the situ-ation. He should keep in remembrance this one fact, to wit: That, owing to the superior brilliancy pertain-ing to the dry mount, little differences in the collar ad-justment are not so perceptible as is the case with bal-samed objects; nevertheless, the difference *is* *there*, and well worth the study necessary to recognize these little differences. Again, should the learner desire to com-pare one objective with another as to their comparative

defining powers, always choose a balsam mount, and the
thinner and weaker the shells (within reasonable limits)
the better and the more palpable will be the comparisons.

We advise every lover of a good objective to provide
himself with a slide of the genuine English podura, as
a matter of course dry mounted, and let him be careful
to keep it dry! A mount that has once, only, been
swamped in water will be probably badly damaged; and
should it occur that water leaks through the cement, it
should be immediately dried by moderate artificial heat,
and subsequently laid away in a warm place for several
hours. Too much care cannot be taken with a really
valuable slide of podura. Prof. Phin says, in his excellent work, "How to Use the Microscope," alluding to
the podura scale, "page after page has been written for
the purpose of showing how the podura *ought to look*,
and still the question seems to be undecided."

A year or two ago a lively discussion sprang up in the
London *Monthly Journal of Microscopy*, between Messrs.
Piggott and Wenham, the former contending that the
true resolution of the podura resulted in dots. Mr.
Wenham, on the contrary, warmly taking sides with
Ross and Beck, holding that the "exclamation points"
were the proper thing to be shown. A month or two
later a most curious article appeared in the "*Popular
Science Monthly*," contributed by Mr. John Michells, and
entitled "The Microscope and its *Mis*interpretations."
The point of Mr. Michell's article was to show the want
of reliability attending observations made with the microscope. To make this point salient, he referred to the

London controversy of the Messrs. Piggott and Wenham on the podura question, and also gave the statements of other English observers. Among the illustrations accompanying this article were five or six cuts showing the same *scale* of Podura as seen in the microscope under different illuminations. Thus one identical scale was exhibited to the reader as widely changed in its aspects as could have occurred had dissimilar scales been selected for the comparison. One or two of the results shown were evidently owing to distortion occurring from the use of a badly corrected objective.

Now let it be remembered that Mr. Michells, in thus stating his authority for the *mis*interpretations of the microscope, worked with legitimate material, and the work over the podura presented was such as had been arrived at by English microscopists of acknowledged competency. There must be something wrong; in what direction shall we look for the cause thereof?

So far as the exhibition of " the markings " of the podura are concerned, they are an easy test; *i. e.*, almost any common-place glass will show them after a fashion, while what would be known as a " real honest working sixth or eighth " will show these markings with considerable force; and such a display would likely be as acceptable to the novice (or to those who had never worked with better objectives), as the view given by a superb high balsam angled glass by Spencer or Tolles. But mark this: Hardly two of the " real honest working glasses," owing to the hap-hazard character of their corrections, will give the same appearances, while the objectives of Spencer and Tolles will have but one and the same constant story to tell. With these, properly manipulated, it is impossible to get aught over the podura but the exclamation points: With these superb glasses such a thing as the reproduction of the several appearances shown by one and the same scale, as exhibited in the cuts by Mr. Michells, becomes an impossibility. And here, in the opinion of the author, may be found the key of the problem, to wit: The better glasses showed the podura tolerably well. The poorer ones, those " real honest working glasses," of which we have heard so much about, were bound to set forth their individual *mis*representations, and thus furnish material for contributions to periodical literature like that of Mr. Michells.

As to the true structure of podura. there can be little doubt; there has never been but one opinion among American observers accustomed to the use of the finest American objectives, and the spines or exclamation marks are accepted as the true resolution of these scales. By the aid of electricity, these spines have been detached from the body of the scale; the detached spine together with the parent scale were photographed and thus presented to the readers of the " Lens."

We have already had occasion to say that Mr. Geo. W. Morehouse, of Wayland, New York, was the first observer who demonstrated the capabilities of the Beck vertical illuminator; when used in conjunction with American object-glasses of high balsam angles,

Mr. Morehouse's results in displaying objects by reflected light under the higher powers of the microscope
were truly astounding, and his paper read before the
Dunkirk Microscopical Society of Dunkirk, arrested
the attention of the author, who immediately repeated
Mr. Morehouse's observations with the liveliest satisfaction. Mr. Morehouse particularly called attention
to the unrivalled views thus given of the podura,
adding that they seemed to settle all question as to the
nature of these markings. Mr. Morehouse observed
also that near the ends of scales the spines were sometimes to be seen projecting beyond the scale itself, and
the author can affirm the same from his experience.
As suggested by Mr. Morehouse the vertical illuminator, from the nature of its action—dealing entirely
with surfaces only—not only deprives Mr. Mitchell's
paper of argumentative force, but fairly turns his
weapons against him, pointing to the podura, indeed,
as a very proper object with which to demonstrate the
reliability of microscope observations.

The student will find the study of the podura a valuable assistance in teaching him the behavior of fine
object-glasses over dry mounts, and when he shall by
dint of long practice become sufficiently expert to detect and appreciate the beauty of the display as given
by balsam aperture objectives as contrasted with the
work of medium angled glasses, he will suffer no regrets as to the time and patience this accomplishment
may have cost him.

We have insisted on the use of the genuine English

podura, and as we have stated, this slide must necessarily be selected by some friend competent to the task. The American podura is quite inferior to the English, yet of the two a slide of the very best American should be accepted rather than a notably poor English specimen. The former will be much better than none at all, but let it be remembered that the English are the desired thing. Now as to the proper appearance of these scales under the objective, we must state that it is perfectly impossible to give any truthful representation on paper. The following cut taken from Prof. Phin's book, copied from the illustrated catalogue of the late Richard Beck, may help matters some, and shows the podura with tolerable precision, when the same is seen with a dry one-eighth, say of 130° aperture magnified 1,300 diameters, and illuminated with light as central as possible. Now let the student take a half sheet of writing paper and roll this up so as to form a tube, and examine the cut attentively, looking through this tube, as is frequently done when examining drawings, etc. Letting a strong light fall from the direction indicated by the shadowed sides of the markings, we think that most of our readers will have no difficulty in noticing that the " exclamation points " seem to " rise up " above (apparently) the general surface of the scale. Our principal design in calling attention to this feature is that he may thus recognize an effect to which we have had occasion to allude to in previous pages, when treating of certain diatom tests. Now this same effect is easily seen in the podura when under the objective, and we have never been entirely able to get rid of it, especially when a glass of moderate angle is employed. Now the reader knows from personal experience just as much of this borrowed cut as does

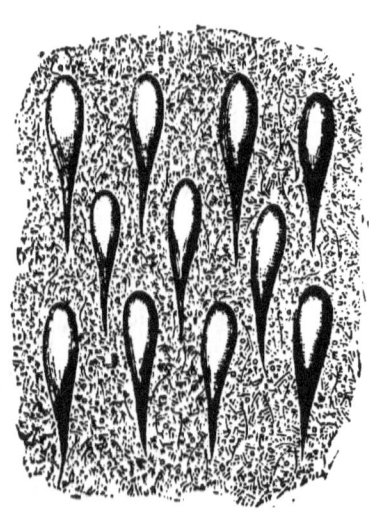

the author. We will, however, hazard the assertion that the illumination shown was about 20° from axis, the glass having aperture as above stated, of perhaps 120° or 130°. Let us imagine then that we thus have such a display under the microscope; and again, that we remove the objective and replace with another of high balsam angle. In the cut on this page we have endeavored to depict the changes.

First it will be seen that with the radial bar in precisely the same position we have the effect of greater obliquity; the shading is more decided and (over the tube) the lights are more intense and brilliant. Now examine this cut with the paper tube; it will be observed that the exclamation points do not "rise" so much as was the former case. They indeed appear in good relief, but they do not seem to float clear from the body of the scale; these appearances are to be watched under the microscope, and the obliquity of the illumination diminished to the least angle without allowing the spines to rise. When the adjusting collar and the illumination are just in the correct position, the small end of the " wedges " or " exclamation points are to be sharply defined. These scales may be examined either in a vertical or horizontal position; as a general thing the vertical will be the preferable position. Most mounts of podura will contain scores of scales, while for the most part there will be less than a half-dozen good ones. The student should look carefully to this, and make the best selection possible. Referring to the use of the vertical illuminator, Mr. Morehouse called my attention to a curious fact connected with the study of these scales, namely, that the smallest and most uninviting of these, as shown by transmitted light, seemed under the vertical illuminator quite as strong, vigorous, and satisfactory as any. We found Mr. Morehouse's observation true as respect to many other insect scales we had occasion to examine.

The following, mounted dry, will form useful objects and may be advantageously studied conjointly with the podura:

Scales of Lepisma saccharina—Degeeria—Macrotoma major—Petrobius maritimus—Pontia brassica—Morphomenelaus—Tinea vestimenti—Gnat—Hipparchia Janira —Wing of Gnat—any of the scales from the lepidoptera will answer a good purpose, and in the summer time the student will find a plenty of these visitors, and from such material many excellent mounts can be prepared with little loss of time and at a slight expense. We recommend that the pupil become well accustomed to handling insect scales before attempting the miscellaneous examination of anything and everything mounted dry that perchance may be directly at hand. After considerable proficiency has been secured, the student may test his powers thus: Let him provide himself with a dry mounted trachea, say of a bee; use the lowest ocular, select the strongest and most vigorous part of the object where the coils are the largest and the strongest, and adjust the glass as nicely as possible thereon. This done, choose one of the longest continuous coils and see *how far* this can be followed towards the smaller end. Now having arrived at the limit, so that the coils can but just be perceived, manipulate the adjustment; now if you succeed in getting very much better definition, the proof is before you that you were in error at the commencement. Make a note of the situation, and throwing the glass intentionally out of adjustment again, repeat the experiment until you shall be enabled to certainly make the initial correction with tolerable certainty and accuracy. This is first-class practice, and will be of the utmost value in general work over dry mounts. The trachea, too, is a real good test object. The author has worked ten hours at a sitting endeavoring to trace these coils to the very last end, but has thus far met with defeat.

As to the general examination of the various dry mounts, or of histological preparations mounted in glycerine, nothing but the most general directions can be given. The previous study of the diatoms will become a valuable assistance to the student, and this study should be still kept up in the examination of dry mounts especially. Of such objects as have palpable size, and can be seen with the naked eye, as is the case with very many histological mountings, the pupil must expect more or less annoyance from the interference of the various shadows, from different parts of the structure, situated in different planes. If, in such a mount it be desired to examine critically minute details, make it a point to select a bit of the object that has perchance been accidentally removed from the main body, and when you prepare your own mounts keep this point in mind. A good general rule with such mounts is to keep the illumination as nearly central as possible. In the adjustment of the objective the same plan is to be adopted, selecting when available the smallest isolated fragment, with the illumination within 35° to 40° from axis (or less) the low angled sub-stage condenser will be valuable. It however requires some study, as we have found by practical experience, to work the condenser so as to secure its best effects. It is a matter, also, that calls for the outlay of time and no little patience. Nevertheless, the reward will repay the effort.

Through the kindness of Mr. Zentmayer, who constructed the apparatus we have been experimenting with condensers of very small diameter, several of these of varying focal lengths were fitted to the same mounting. Either glass could thus be used at pleasure. The mounting was conical and the base being also of small diameter, say about one-fifth of an inch, our idea was by this arrangement to keep the angle of the condensing lens as low as possible, and by the peculiar form of its mounting to illuminate as little as possible of the object and with an extremely acute cone of light and at the same time, allowing sufficient lateral range that the instrument might be placed in as oblique position as possible. The idea seemed theoretically sound, and we have spent much time in the experiments hoping to give our readers something of value, but we were doomed to disappointment. (Not the first experiment of ours which has failed). The instrument is very troublesome to use, requiring often some little time to coax the light through the little cone, and when once obtained, the slightest movement of the mirror will destroy what has been done. Nor when doing its best were the results any better than those obtained by the Gundlach, or Beck cheap inch which has already been recommended for general condensing purposes. We still believe that some improvement will be made in the direction of our experiment, and hope that other observers will make similar efforts until further success shall be secured.

OIL IMMERSION OBJECT-GLASSES.

Since the introduction of the duplex objectives another form of four-system, wide apertured glass, constructed by Carl Zeiss, of Jena, Denmark, has made its appearance in this country.

With the advent of these glasses came also the announcement that their balsam apertures excelled those of the American duplex. A novel feature in the mounting of the oil immersions was that they were not provided with adjustment collar; it being asserted that the only requirements necessary to secure the perfect action of the object-glass were to use the same with a tube of exactly ten inches length, and to employ *cedar oil* as the immersion medium.

It was further stated that when these simple requirements were alone attended to, that no further adjustment of the objective would be required, and that the performance of the object-glass would be found unequaled.

As might be supposed, the above claims on behalf of the oil immersions were destined to receive attention at the hands of American microscopists.

A *fact* well known to experts concerning wide apertured objectives is this, viz: An object-glass exactly corrected to the eye of A may not be in exact correction to the eye of B without suitable change of the collar adjustment.

During a sitting with Mr. Chas. A. Spencer at Buffalo last fall, the above fact became immediately apparent. Mr. Spencer's adjustment of his duplex one-tenth objective (which we were then using) being *invariably* three divisions of the collar graduations nearer "closed" than my own.

It is obvious therefore that an " oil immersion " suitably corrected to the vision of Mr. Zeiss when worked with ten-inch tube, might require a special correction when in the hands of Mr. Spencer, or myself. Feeling therefore assured of the error embodied in the *popular* representations concerning the oil immersions of Mr. Zeiss, the author solicited and obtained one of Mr. Zeiss' circulars, which, being official, will, it is believed, be read with interest. With the exception that a few typographical errors have been corrected, the following is otherwise a *verbatim* copy.

NEW OIL IMMERSION OBJECTIVES.   ONE-EIGHTH INCH. ONE-FIFTH INCH.

" This object-glass is adapted to an immersion fluid, which in refraction and dispersion is equal as nearly as possible to the crown glass, a plan proposed by Mr. J. W. Stephenson, of London, as a devise for increasing the aperture and for dispensing with correction.

"The lens constructed on this plan—a four-fold system of pp. one-eighth focal length, calculated by Prof. Abbe—shows an aperture of unusual amount, combining a most perfect definition with a reasonable working distance. It works equally well through thick, thin, and thick covering glass without needing special correction, owing to the identic refraction of the immersion-fluid and the covering-glass.

"The numerical aperture of the object-glass (according to Prof. Abbe's definition, the product of the sine of angular *semi* aperture with the refractive index of the medium exposed in front) is brought to the number 1.25 exactly, which corresponds to a balsam angle of 113° and is in the ratio of 5:4 greater than would be the maximum aperture, 180° in air, of a dry lens, considered in its numerical equivalent, the resolving power correspondingly affords a visible increase compared with immersion-lenses of the common system, which generally do not exceed 1.10 in the numerical equivalent of aperture.

"As to the immersion-fluid a large number of experiences has shown the oil of cedarwood (ol. ligni cedri) to be the most fitted, though it is in a slight degree still less refractive than ordinary crown-glass; other liquids of higher refraction exceeding too much in dispersive power. This oil (of which a sample will be sent with the objective) may be got anywhere in sufficient purity.

"For controlling its qualities or those of other liquids, which perhaps might appear convenient, a special test-bottle is forwarded with parallel plane faces and a prism of crown-glass, cemented to the stop of this bottle. The vertical spar of a window seen through the prism and through the oil beneath, *should appear without a sensible deffection and should show bodies only slightly colored,* if the liquid has the right quality.

"The pure oil, ligni cedri, will afford the best color

correction for *oblique* light; for observation with central light the image may be obtained still more colorless if a suitable quantity of a higher dispersive oil (oil of fennel seeds or anise oil) be mixed with that of cedar-wood. But in this case the refraction of the mixture should be reduced to the original refraction of the oil of cedar-wood by adding some pure olive oil; the test-bottle being always applied for regulating the mixture.

" A few other oils: The oil of copaiva-balsam and the oil of sandal-wood approach so close to the oil of cedar-wood in refraction and dispersion, that they may be used for it, instead of it, if they should be preferable in any respect, provided the test-bottle has stated the right quality of the sample.

" The use of these oils will not be injurious to preparations which are hermetically closed, as it is likewise necessary for water-immersion, although the black varnish on some slides (not on all) will be slightly dissolved by a prolonged exposure to the oil, its action will not be offensive within a moderate time; and in most cases any contact of the fluid with the edge of the covering-glass may be avoided, by the oil not being applied in a greater quantity than is necessary. A minimum drop on the covering-glass, and such a one on the front lens being quite sufficient for observation; besides that, the varnish of the slides may be perfectly secured against the oil by a solution of shellac in alcohol.

" A special advantage will result from this mode of immersion for the more difficult dominions of petrographic work; since rock-slices for inspection with the microscope will become perfectly transparent, using the oil without needing a polished surface or a covering-glass cemented on, and may be observed to a greater depth than would be accessible to the higher powers of the ordinary system.

"In every case the new object-glass, not considering the greater optical capacity in bringing out difficult structures, will prove exceedingly convenient for use, from dispensing with any trouble in finding the right correction.

"Of course the full performance of the increased aperture can be effective only on preparations which are mounted in balsam (or any other medium exceeding 1.25 in the refractive index) or which, if mounted dry, perfectly adhere to *the covering-glass*. On objects separated by air from the covering-glass the lens will not work better than any good immersion-objective with an aperture equivalent to an air-angle of 180°.

"Besides this, for displaying the full performance in oblique light, the illuminating-apparatus must yield pencils of greater obliquity than are directly accessible to a slide from air. The most simple devise for getting light on such an obliquity without needing a special apparatus (immersion condenser) is a plano-convex lens cemented to the under surface of the slide by a minimum quantity of oil. The ordinary mirror of the microscope brought to a moderate distance from the axis will now yield pencils of any wanted obliquity; a lens fit for this use will be added to the object-glass.

" If ordered for English microscopes the lens will be perfectly corrected for a tube of ten inches *exactly*, and no sensible deviation from this length would be admissable without a loss of perfect definition. However, owing to the slight defect of refractive power in the oil of cedar-wood, some advantage may be found *by lengthening the tube for one-half to one inch while observing through extremely thin covers (less than 0.004 inch), and by shortening it for one-half to one inch in the case of very thick covers (exceeding 0.008 inch).* *

" The object-glass is made with fixed brass work and with standard screw (like all my object-glasses); the price of it is 240 marks; the price of the one-twelfth is 320 marks; the aperture guaranteed to be not less than it is stated above. The lenses are screwed together with moderate pressure and may be unscrewed without great effort, but I caution expressly against unscrewing them; owing to the great aperture the system is extremely sensible to the slightest defect of centering the smallest particle of dust, or the least moisture getting into the screws, and the unavoidable difference of pressure when screwing the lenses together, would cause a sensible loss in the performance of the glass."

JENA, March, 1878.   (Signed)   CARL ZEISS.

One can hardly read the foregoing somewhat contradictory document without arriving at the conviction that Mr. Zeiss has suffered severely at the hand of his translator. Be this as it may it will be noticed that he asserts that the oil-immersions practically NEED adjust-

---

* Italics mine, J. E. S.

ment, which adjustment he recommends be accomplished by manipulating the draw-tube to a possible extent of *two inches* when working over covers varying from 0.004 to 0.008; and here let the reader be reminded that these figures do not by any means express the maximum variations of cover-glasses.

I have had an opportunity of critically examining but one of these objectives, a one-eighth, imported by a friend. This glass when worked with ten-inch tube over my usual covering-glasses required downright engineering of the draw-tube; furthermore, when worked over covers 0.008 inch thick, it was impossible to correct the glass at all, as the draw-tube of my stand when fully closed would not decrease the distance sufficiently to secure the proper correction of the objective.

The performance of this one-eighth (when exactly corrected) was very fine indeed, and closely resembling the work of the Tolles or Spencer duplex. Superb as its definition was, it did not excel that of my Spencer duplex one-fourth—a glass of half its nominal power. The balsam aperture of the one-eighth was about the same as that of my duplex glasses.

During my first evenings with the oil immersion the cedar *oil* behaved very well, but on the second and third sittings it made me trouble enough. With the tube inclined (as usual) the oil would most unaccountably take a notion to run away; when this occurred, any attempt to reinforce by adding more of the oil, made things worse; the only remedy was to clean the slide nicely and commence operations again, *de novo*.

The author learns from good authority that Mr. Zeiss has recently extended the balsam aperture of his oil-immersions improving thereby the performance.

Oil-immersions are now made by Mr. Tolles and Mr. Spencer. These makers, however, supply the collar adjustment (as should have been done by Mr. Zeiss). Mr. Spencer's objectives we are informed are arranged so as to work with *oil* or glycerine contact.

While in the oil business as above stated, it occurred to me to try the effect of various oils with my duplex glasses. After a little experimenting, I found that ordinary *kerosene*, well washed in alcohol, worked most beautifully with both the Tolles one-tenth and the Spencer duplex one-fourth, with either glass when thus immersed I got the most exquisite displays of the most difficult known tests. It thus came to the surface that I had harbored oil immersions while ignorant of the fact.

That the cedar oil *diminishes* the *range* required in process of adjusting for varying thickness of cover must be granted, hence the novice using the Zeiss immersions being necessarily restricted to this more limited range might possibly get better initial performance than would result in his first attempts to manipulate a Tolles or Spencer four-system objective.

In dispensing with the cover adjustment and its accompanying mechanism (such as furnished by Spencer or Tolles), falling back on the clumsy draw-tube as a substitute, the cost of construction is *materially* reduced. Those affected with the *res angusta domi* might doubtless contract with either Tolles or Spencer for objectives on the model of the Zeiss, and fully equal to the latter in performance, and at figures considerably below those quoted in the Danish catalogue.

Our experience with the Zeiss oil immersions has neither weakened our appreciation of the screw-collar adjustment, nor our high estimation of wide apertured objectives of American manufacture.

CHAPTER VIII.

A WORD OR TWO ON VOLUMETRIC ANALYSIS.

The value of the microscope to the medical profession is greatly enhanced by the conjoint use of a little chemistry; in the examination of urinary deposits it will often occur that neither the microscopical nor the chemical analysis *per se* would be complete or entirely satisfactory, while the combined results of these might lead directly to the information desired.

Micro-chemical examinations of urines are of the utmost importance to the practitioner; among the best English works on the subject may be named those of Golding Bird, Beale, Roberts and Harley. The beginner will derive much satisfaction from the work of Dr. Bird, which although written twenty years ago is essentially good at this day. The reader should keep in mind in the perusal of either of the above authors, that microscopy has of late years suffered material advancement, hence it is with the superior objectives now at our command, we are able to cross-question some of the plates contained in the above-named books.

Almost anyone looking over the representations of " tube-casts " as pictured by Dr. Beale, would naturally arrive at the conclusion that there could be no doubt as to the recognition of these objects in practice. It becomes my duty to say, therefore, that in neither of the works referred to, can be found a reliable representation of a genuine "tube-cast"; while on the other hand the learner depending entirely on the information conveyed to the eye by the plates is surely liable to be misled. It is very far from our purpose to make any attack on the English authorities mentioned; from either of them we have derived a great deal of useful and valuable information, and they are to-day works, to which we make frequent reference. And then again we doubt seriously, if an accurate idea of a "pale hyaline cast" *can* be conveyed by means of drawings. The learner, however, may rely on the *text* of our favorite authors, and at the same time use due care that not every adventitious filament, hair, etc., be accepted as a genuine "tube-cast."

To detect with certainty a genuine "cast," one of the feebler order requires just as much care and study as would be required to show the lines on the Nos. 18 or 19 of the Moller test plate, and requires also just as good instrumentation.

Having thus alluded to one serious source of error, we have to say that it is no part of our purpose here to write a treatise on urinary deposits, although we hope to accomplish something in this line at no distant day our present purpose is to give the medic a hint or two as to the little apparatus required in the volumetric analysis of urines, and to instruct him in the preparation of the necessary chemical solutions employed in such analysis; $15.00 or even less will purchase a tolerably effective outfit.

APPARATUS.

Six test tubes; let four of these be large size. Spirit lamp. Four or five two-ounce beakers. Bink's 250 grains burette. Two or three small glass funnels, say of two-ounce capacity. Filters to match the above. Two or three glass rods. Two or three brushes for cleaning test tubes; ditto, for burette. Litmus paper, blue and red. Two measuring pipettes, one-half, one and two fluid ounces. Urinometers.

All of the above can be obtained of Messrs. G. Tieman & Co., 67 Chatham street, New York, and will cost less than $7.00

Additional to this list we recommend the purchase of three or four " Marais " graduated tubes for approximative analysis. These cost $1.50 each. At least *one* of these should be provided.

A delicate balance, one turning with one-fourth grain when loaded with an ounce in each pan, will be required. Most of the books treating on volumetric analysis demand that the balance should be quite an expensive one. Nevertheless, the simple little models which can be procured almost anywhere, and costing less than $3.00, can be made to answer tolerably well. Those to whom the expense is not objectionable will find a really fine balance enclosed in a glass case a luxury. Ours, made by Troemner, of Philadelphia, turning with, say one-fiftieth of a grain, costing $40.00, has given excellent satisfaction. Reliable instruments, by Becker, can be obtained in the cities, at prices from $10.00 upwards. Those who have used them speak highly of them; but for the purposes of the practitioner, such instruments are not strictly a necessity. The small German scales found in the office of almost any physician can be made, with a little attention, to do very fair work. If such be selected, it will be of the very first importance to possess a reliable set of weights; those furnished with the cheap instruments are not to be tolerated. Nor is it safe to appeal to the nearest druggist. The safest method is to make one's own, which is accomplished with no great outlay of time or money.

First, purchase a set of Troemner's "aluminum" grain weights, costing fifty cents. These consist of a five, four, *two* three's, two, one, and one-half grain weights. With these in hand it will be easy to construct an accurate set. Specific directions are unnecessary.

A thousand grain bottle will also be wanted. This can be purchased in the shape of a regular "specific gravity bottle;" its cost is $2.00. The practitioner can readily adapt an ordinary bottle to the purpose, simply selecting one of such capacity that when a volume of one thousand grains of rainwater at 60° F. is placed therein, the water shall rise part way in the neck, and its place marked with a file, we have on hand some half-dozen 1,000 grain bottles, which we have picked up from time to time, and which are in regular use, the specific gravity bottle being held in reserve as a standard, or for the determination at times of the density of liquids.

Imprimis.—Let me recommend to the practitioner that he eschew the entire system of weights as recognized by the profession, and confine himself to that one simple unit—the grain. This, in the case of chemicals, can be adopted with perfect ease, and much confusion thus avoided. In the following pages no other weight will be mentioned. Referring to liquid measures, the case would be somewhat modified, and we shall have occasion probably to speak of fluid ounces, drachms, etc. Now there are two fluid ounces in vogue; the one is known as the English " imperial," and is equivalent to a volume of distilled water at 60° F., weighing 437.5 grains. This, also, is the ounce avoirdupois. The United States standard fluid ounce is quite another thing. This is equal to a volume of distilled water at 60° F., weighing 455.69 grains, sixteen such ounces making one pint. The United States ounce will be the one referred to in this book, when speaking of fluid ounces and drachms, eight of the latter being equal to one of the former. The practitioner should keep in mind these differences between the United States and the English fluid ounce, when reading English authors.

In the selection of his chemicals for the preparation of the standard solutions, the utmost attention must be paid to their purity. They should not be bought at the " nearest drug store." It will be always advisable to purchase them from leading dealers. The formulas which will be given relate only to pure chemicals. Due attention also should be given to the manipulations, that the chemicals shall not become contaminated. Even after the standard solutions are successfully prepared, the careless operator may ruin a solution by thoughtlessly pouring the remaining contents of the burette (after an analysis) into the wrong bottle. The same glass rod ought not to be used in different solutions without being cleansed. Every bottle should have its own stopper, and these should not become interchanged. Every piece of apparatus used in an analysis should, at the termination thereof, be immediately cleansed.

The practitioner will find it convenient to make his several standard solutions in quantities of 3,000 grains. With the exception of the one used for the determination of sugar, they all have reliable keeping properties, and as 3,000 grains can be made with about as little trouble as 1,000, it is an economy of time to do so.

Procure, if possible, suitable bottles for the standard solutions, furnished with a "pouring lip." These are so much handier in filling the burette. Ours were obtained from an ink manufactory, and answer the purpose perfectly.

Analysis for Urea — Standard Solution. — Weigh thirty-eight and six-tenths grains of pure red oxide of mercury and place the same in a large test tube, add a little nitric acid, c. p., and apply the heat from the spirit lamp. The oxide will appear to crust and little inclined to dissolve. By keeping up an uniform heat, meanwhile stirring with a glass rod, the oxide will become gradually dissolved. Should it, however, become necessary, add a little more acid carefully, little by little, maintaining the gentle heat and stirring with the rod until all of the oxide shall be dissolved, the object being to use the *least* amount of acid possible. The process being completed, pour the whole into a thousand grain bottle, rinse two or three times with distilled water, adding this to the measuring bottle, and finally add distilled water to make the whole to 1,000 grains. Let it stand for a few hours, after which time should there form a precipitate of the basic salt, add two or three drops more of the acid. The solution may now be transferred to the regular bottle for use. Two hundred grains of this standard solution are (should be) equivalent to one grain of urea.

Baryta Solution.—In one bottle make a *cold* solution saturated with nitrate of baryta; four ounces of distilled water will be sufficient. In another bottle saturate eight ounces of distilled water (cold) with *caustic* baryta. When the solutions are fully saturated, which may be known by the baryta remaining in excess in the two bottles, allow a little time for the two solutions to become clear; then carefully decant as much as possible of the nitrate solution into your regular bottle, and to this add *twice* the volume of the caustic solution, the two combined forming the baryta solution for use in analysis.

Carbonate of Soda Paper.—In a bottle prepare a saturated solution of carbonate of soda. When fully saturated pour into a large dish or platter, provide ` sheets of ordinary printing paper, saturate these and suspend the sheets until quite dry; cut into strips one inch wide by six or seven inches long, and preserve in a wide mouthed bottle fitted with a well-fitting stopper.

Measuring Bottle.—Procure one of the long and slim

four-drachm bottles found at almost any druggist's, say with an interior calibre of one-fourth inch. Mark this with a file at heights corresponding to 50, 66, 100 and 200 grains. This is easily accomplished with the aid of the burette. The plain bottles without a neck are to be preferred.

The Analysis for Urea.—Take a sufficient quantity of the urine, and if albumen is present clear it from the same; next, pour into the measuring bottle to the 100 grain mark. Transferring this to a common wineglass, in a like manner measure 50 grains of the baryta solution; add this to the other in the wine-glass, pour the whole on a dry filter, receiving the filtrate in another glass. Should the liquid come through clear, it is well; if not, it must be refiltered until it does. Three or four filtrations at the most will generally accomplish the desired end. While the filtering is in operation fill the burette to the " 0 " mark with the standard solution.

With the measuring pipette transfer one-half drachm of the clear filtrate to a clean wineglass, adding half its quantity of distilled water. Have in readiness a strip of the carbonate of soda paper and a glass rod. Deliver the standard solution from the burette to the filtrate as long as any precipitate is distinctly seen to form, stir with the rod, and place a drop from the glass in contact with the paper, waiting a moment to observe the reaction. If the paper continues white, add again from the burette, and stir again, placing a second drop on the paper, and thus continue carefully until the drop transferred to the paper strikes a yellow color. Now look among the previous drops on the paper and see if there are any indications of the yellow. If so, you have used too much haste, and the analysis must be repeated. If to the contrary the yellow is only to be observed at the last test made on the paper, the analysis is ended. Read the burette, multiply this number by 12 ÷ 100 and you will have the number of grains to the fluidounce of urine.

It will be well, before making a practical use of the standard solution, to test the same. Proceed thus:

Procure, say one-fourth ounce of pure urea. This can be had of the leading druggists at a cost of about twenty-five cents. Of this weigh carefully two grains, which dissolve in 200 grains of water. Take 50 grains of the solution and tritate as above. When the carb. of soda paper strikes a yellow color, desist and read the burette. Now the 50 grains of solution contained, of course, one-half grain of urea, and the burette should, if the standard solution have been properly prepared, read at the 100 mark. If the reading should be less than the 100, the solution is too strong, and must be weakened by the further addition of distilled water. If, on the contrary, the reading should be more than 100, the solution is too weak, and more of the mercury must be prepared and added. For the purposes of the practitioner there is no necessity of being over precise. If the burette shows but a small error, plus or minus, it will be sufficient to note the fact on the label of the bottle, applying a proper correction to the future analyses. In the examination of the reactions on the carb. of soda paper, especially in the evening, we find a hand-magnifier of much service.

Chloride of Sodium — Standard Solution.—Dissolve forty-four grains pure nitrate of silver in 3,000 grains of distilled water; employ a clear white glass bottle and set the same in the sunlight for eight or ten hours. A dark brown or black precipitate will probably form; when this settles to the bottom, filter into a clean bottle, decanting carefully so as not to disturb the sediment; 200 grains of this standard solution should equal one grain of chloride of sodium. Before being put to practical use the solution must be tested.

Solution of Chromate of Potash is made by saturating three or four ounces of distilled water; an excess of potash remaining undissolved is not objectionable.

To Test the Standard Solution..— Procure a nice clean lump of " rock salt," crush, and select of the cleanest, ten or fifteen grains; powder roughly and dry with care, using but gentle heat. When thoroughly dry, dissolve two grains in two hundred grains of distilled water, and of this measure fifty grains into a wine-glass, add a little water and one or two drops of the potash solution. Fill the burette and tritate. The addition of the first drops of silver to the salt solution may be followed by the appearance of a red precipitate, but on stirring with the glass rod this will redissolve; continue the addition of the silver little by little, stirring well after each addition, until the further delivery of the silver is followed by a *permanent* precipitate imparting a red color to the contents of the wine-glass. This finishes the test, and if the standard solution is of the proper strength, the burette will read at the one hundredth mark; if too strong or too weak, add silver or water as the case may demand, and repeat the test. The indications of the burette furnish a ready guide as to the amout of correction necessary, hence the second test ought to be sufficient.

Analysis of Urine for Chloride of Sodium.—Filter say one-half ounce of the urine, and with the measuring pipette introduce one fluid drachm of the filtered urine into a wine-glass, and add three volumes of distilled water, and also two or three drops of the saturated potash solution. Now with a bit of litmus paper test the reaction of the mixture; it should be rendered faintly alkaline by adding carbonate of soda, or dilute nitric acid as may be found necessary. Fill the burette and tritate according to the directions given above, until the permanent precipitate makes its appearance. Now read the burette, divide this number by twenty-five which will give you the number of grains of chloride of sodium per fluid ounce of urine.

Should the specimen of urine be very high colored it may be somewhat difficult to detect the first appearance of the permanent precipitate, hence the accuracy of the analysis will be impaired. In such cases, therefore, it is better first to decolorize the urine; this is tolerably well accomplished by adding a drop or two of a solution of permanganate of potash, and with the spirit-lamp bring the mixture nearly to the boiling point. A brown precipitate will be observed which is to be removed by filtration, and the filtrate used for the analysis according to the preceding directions.

Analysis for Chlorides Approximately.— Fill a Marais graduated tube to the ounce mark with filtered urine, make strongly acid with nitric acid; provide a solution of nitrate of silver, forty grains to the ounce of water; add an excess of this to the urine in the tube, allowing the whole to remain quiet for twenty-four hours. At the expiration of this interval the volume of precipitate may be read from the graduated scale, each .2c.c. of precipitate will equal 0.19 grains of chloride of sodium; in the above approximate analysis the acid must not be omitted, otherwise a precipitate of phosphate of silver might vitiate the results.

Analysis for Phosphoric acid.—Dry carefully a sufficient quantity of chemically pure nitrate of uranium, using gentle heat; of this take 106½ grains and add 3,000 grains of distilled water; 200 grains of this standard solution equals one grain of phosphoric acid.

Solution of Acetate of Soda.—Four hundred grains of acetate of soda are dissolved in six fluid ounces of water, and to the solution add 800 grains of acetic acid. The commercial article known as "No. 8" will answer.

Solution of Ferro-cyanide of Potassium is made by dissolving one part of the salt in ten parts of water.

Test Solution.—Dissolve 50.4 grains of phosphate of soda in 1,000 grains of water; 100 grains of this solution equals one grain of phosphoric acid.

To Test the Standard Solution.—Measure fifty grains of the above *test* solution into a beaker; add one-fourth volume of the acetate of soda solution, increase the volume of this mixture three-fold by the addition of water; fill the burette with standard solution, and have in readiness a white plate or saucer, on which have been placed several separate drops of the ferro-cyanide. Heat the contents of the beaker over the spirit-lamp, and keeping the same tolerably warm, tritate with caution, and when a drop from the beaker on being placed in contact with the ferro-cyanide strikes a brown color, the analysis is ended; and if the standard solution is of the proper strength, the burette will read at the 100 mark. Any error of standard solution noticed is to be corrected and the analysis re-peated with the standard solution as amended.

Analysis of Urine for Phosphoric acid.— With the pipette measure two fluid drachms of the urine into a beaker, and to this add one-half drachm of the acetate of soda solution, also an extra drachm or so of water; have ready the white plate and the ferro-cyanide solution; warm the contents of the beaker over the spirit-lamp, and maintaining the heat, tritate as in the above test; continue the addition of the standard solu-tion until the drop from the beaker when in contact with a drop of the ferro-cyanide shall strike a brown color, which terminates the analysis. Read the burette, the number shown divided by fifty will indicate the number of grains of phosphoric acid to the fluid ounce of urine.

Analysis with the Marais Approximate Tubes.— The tube is to be filled to the ounce mark with filtered urine, add also two drachms of the acetate of soda solution, afterwards add an excess of a solution of nitrate of uranium (one to ten); placing the thumb over the end of the tube; mix thoroughly and allow the tube to remain quiet for twenty-four hours, after which time the volume of precipitate may be read off, ever 2cc —.016 grains of phosphoric acid.

Analysis for Earthy Phosphates, using the Marais Tubes.—Fill the tube to the ounce mark with filtered urine as above, add an excess of strong liquor ammonia; mix well and set aside for twenty-four hours, at the end of this interval note the volume of precipitate; each 2cc will correspond to .06 grains of the earthy phosphates.

Analysis for Sulphuric acid Standard Solution.—In 3,000 grains of distilled water, dissolve forty-five and three-fourth grains of pure chloride of barium; 200 grains of this standard solution are equivalent to one grain of sulphuric acid. If the solution be prepared with due care, it may be used without special testing as it cannot be very well tritated.

Solution of Sulphate of Soda.—One part of soda to ten of water.

Analysis of Urine for Sulphuric acid.—With a pipette measure two fluid drachms of urine into a beaker, dilute this with twice as much distilled water, adding also two or three drops of hydrochloric acid; bring the beaker over the spirit-lamp until the contents become hot, fill the burette with the standard solution; deliver from the burette into the beaker a few drops which will cause an instant precipitate of the sulphate of baryta; this will gradually sink to the bottom of the beaker; continue thus the addition until no further precipitate can be detected by the eye. By waiting a few moments the liquid above the precipitate will become clear, when another drop or two of the standard solution may be added. Should this cause a further precipitate, continue in like manner to add the standard solution; when no further precipitate forms, place a few drops of the sulphate of soda solution in one of the smallest test tubes, and to this add a drop or two from the beaker; if a white precipitate appear in the test tube, the standard solution has been added to the beaker in excess; and if the precipitate be very dense, the analysis will have to be repeated, using greater care towards the latter part. The number of grains finally shown by the burette, divided by fifty, indicates the number of grains of sulphuric acid to the fluid ounce of urine.

The final determination in this analysis as to the precipitation of all the sulphuric acid renders this tritation somewhat more troublesome to the beginner than the preceding. The practitioner should therefore, by repeated trials, become acquainted with the details. In examining the test tube as to the precipitate, it should be held to strong light, and a hand magnifier we have found to be of material assistance.

Analysis for Sugar—Standard Solution.—Dissolve 51.98 grains of pure sulphate of copper in 500 grains of distilled water. Keep this in a bottle by itself.

Dissolve caustic soda in distilled water until the specific gravity becomes 1.12, or the specific gravity may be determined by the use of Baume's hydrometer, which should float at 16½°. To *each* 1,000 grains of this solution add 259.90 grains of pure *crystalized* Rochelle salts. This constitutes the caustic solution.

One volume of the copper solution mixed with *two* volumes of the caustic solution forms *Fehling's Standard Solution*, 200 grains of which are equivalent to one grain of sugar.

These solutions must be kept separate until wanted for use, mixing only the quantity required from time to time. The copper solution is quite stable, but the other is liable to deteriorate by keeping. It will be advisable for the practitioner to use the caustic solution from an ounce bottle fitted with a tight stopper, which can be refilled from time to time, the stock bottle being kept in a cool, dark place. On mixing the two together a precipitate forms which will immediately disappear on shaking. Should, on boiling the same, the liquid retain a clear blue color, it is in good condition. Heat should always thus be applied in advance. By complying with the directions given, the caustic solution can be kept in order for years; and it is well to make enough at once to last for at least a year, and the physician is reminded that this Fehling's test for sugar is daily in demand.

Test of Standard Solution for Sugar.—Dissolve four grains of pure grape sugar in 400 grains of water. This solution should always be used when freshly made. In the preceding analysis it will be observed that the burette has been filled with the *standard* solution. Now

in the operation for sugar this is not the case. Therefore fill the burette to the "0" mark with the graduated
solution of grape sugar. Into the measuring bottle
before described measure 66 grains of the copper solution, the proper height being marked on the bottle with
a file. Add of the copper solution to that in the measuring bottle to make the volume equal to 100 grains.
We then have 100 grains of the standard solution mixed
for use, and which are equivalent to ½ grain of sugar.
Pour the standard solution from the measuring bottle
into the largest test tube; rinse the former with a little
water, and add this to that in the test tube, bringing
the latter over the spirit lamp; heat to the boiling point,
and notice, first, if the standard solution retains its
clear blue color; if so, proceed to tritate, adding from
the burette a few drops at a time, bringing the mixture, after each addition, to the boiling point. Continue thus, adding from the burette until the blue
color of the mixture in the test tube shall have almost
entirely disappeared. At this stage of the decomposition of the copper the mixture in the tube will
have become thick and of a greenish yellow color, and
will show but little disposition to settle clear. Now
add cautiously a few drops from the burette; bring to
the boiling point again, and, holding the tube in an inclined position near a sheet of white paper, observe the
thin edge of the liquid as to color. Should there appear
a trace of the blue yet, continue the addition from the
burette, and also the heat, until the last trace of the
blue shall have disappeared. Resting the tube in a ver-

tical position, wait a moment or two, and if the reduction of the copper be complete, the sediment will sink in a very short interval of time, to the bottom of the tube; and if care has been taken not to overdo the thing, *i. e.*, introducing more than was necessary from the burette, the analysis is completed and the burette may be read.

Now the 100 grains of the standard solution in the test tube would require ½ grain of sugar for its reduction, and 50 grains of the solution in the burette will contain ½ grain of sugar. The burette should therefore be read at 50°. The student is advised to repeat the operation several times until he shall have become familiar with the reactions. The quantity of test solution is ample for this. If, after satisfactory trials, it shall be found that the standard solution is to strong or too weak, water or copper may be added to the copper solution without making any change in that of the caustic, and the test should again be repeated.

Analysis of Urine for Sugar.—Prepare the standard solution for use just as has been directed; *i. e.*, measuring 66 grains of the caustic solution, and adding the copper solution to make the volume of 100 grains. Pour, as before, into the largest test tube, rinsing the measuring bottle, adding the "wash" to the contents of the tube. A very little clear water may also be added to the tube. Fill the burette to the " 0 " mark *with the urine to be analyzed*, bring the test tube over the spirit lamp, and heat to ebulition. Now go on and tritate just as directed in the preceding test, observing due care, when near the complete decomposition of the copper. When this shall be obtained, and the blue color of the standard solution has disappeared, the sediment in the tube being also inclined to settle quickly, the burette may be read.

Now, to determine the amount of sugar from the reading of the burette is a simple question of the " rule of three," suppose that the number marked by the burette to have been 228, then it follows that 228 grains of the urine contained $\frac{1}{2}$ grain of sugar; Hence 456 (one ounce) of the urine would contain 1 grain of sugar, or, as we would say, " one grain per ounce." It may be well to suggest to those who have allowed their mathematics to become a little rusty, that even in solving proportions, simple as are the computations, that it is necessary to look after one's decimal points. Thus, in the above example, the expression would be — 456 $\times$ 0.50, $\div$ 228 $=$ unity.

Use of the standard solution in the ordinary testing for sugar. The practitioner will very often have occasion to simply test for the *presence* of sugar, when the exact amount is not necessary. In fact, after becoming thoroughly familiar with the behavior of the chemicals, he will be enabled to give a close guess as to quantity. The urine should in all cases be cleared from albumen. To merely detect the presence of sugar it is only necessary to pour a dozen drops or so of the copper solution into one of the largest test tubes (these are always better when there is boiling to be done), and to add twice the volume of the caustic solution. Bring to the boiling point, and add urine to the tube directly, continuing such additions and bringing to the boiling point substantially as in performing the regular analysis. If, when the amount of urine thus introduced shall be equal to the original volume of the test solution, and there be no change of color, it may safely be assumed that sugar is not present.

Even when the regular analysis is contemplated, the preliminary trial test should be instituted. For instance, we do not care to attempt the regular analysis, when by the rough test we are assured that there is no sugar in the specimen. Again, when, by the trial test, it shall be found that sugar is present, and in large amount, then it may sometimes be better to dilute the urine with an equal, or even six times its volume of water; of course, allowing for this reduction in the computations after reading the burette; the accuracy of the analysis is enhanced by the dilution of the urine. It may be further remarked that different specimens behave somewhat differently. The directions given will, however, be found ample, and the practitioner having frequent use for the sugar test in a short time becomes perfectly at ease with the manipulations.

Analysis for Albumen.—The volumetric method, one of "trial and error," involving several filtrations, is somewhat tedious; too much so for the practical purposes of the medical practitioner. The author was hence induced to experiment with the "Marais" approximate tubes comparing results with those obtained by two of the regular methods. These were so far sat-

isfactory as to lead to the employment of the approximate tubes in his general practice; and the degree of
accuracy afforded by the use of these tubes is quite equal
to the ordinary demands of the medical profession.
The principal source of error arises, as we believe, from
the fact that the coagulated albumen will at times pack
closer in the tube than at others. By using three or
four tubes in one and the same analysis, taking the
mean of the results, the approximate process will, in
most instances, be all that can be desired. Proceed as
follows:

In a *clean* evaporating dish, or a Florence flask, coagulate by heat three fluid ounces of the urine to be
tested; set aside until nearly cold; shake well; now
take three marais tubes, fill the first full, also the
second with the treated urine and coagulum; pour the
remainder into the third tube, rinsing the dish with a
little water, adding the wash to the several tubes. The
three tubes can be thus made to contain the entire
coagulum from the three ounces of urine; set aside for
twenty-four hours, after which time read the several
tubes; add the readings together and divide by three,
and every two *whole* cubic centimetres will represent
one grain of albumen to the ounce of urine.

In the daily routine it will often suffice to use but
one tube, in which case all that is necessary will be to
coagulate a single fluid ounce of the urine; when cool,
shake, and pour into the approximate tube, rinsing the
dish as before, adding the wash to the tube; at the end
of twenty-four hours read off the amount of coagulum;

every two *whole* c. c. will equal one grain of albumen to the ounce of albumen.

Reaction.—Urines being at times either acid, neutral, or alkaline, it is often interesting, not only to observe as to the fact, but also as to the *degree* of acidity or alkalinity; to accomplish this prepare the following solutions:

Test Solution for Acidity.—In 1,000 grains of distilled water dissolve ten grains of pure hydrate of soda.

Test Solution for Alkalinity.—In 1,000 grains of distilled water dissolve 15.75 grains of pure oxalic acid. Equal volumes of these two solutions will exactly neutralize each other.

To Test the Degree of Acidity in a Sample of Urine. —Fill the burette with the soda solution; measure one-half fluid ounce of the urine into a wine-glass; deliver the soda from the burette into this a few drops at a time, stirring well after each addition and testing with litmus paper; when the mixture fails to affect the latter, read the burette twice, the figures read will indi- cate the number of grains of the solution employed required to neutralize a fluid ounce of the urine. To test for alkalinity proceed as above, using the acid solu- tion in place of the soda, and also red litmus paper in place of the blue. The daily variations in any particu- lar case can in this manner be determined and recorded.

Proportion per Fluid Ounce of certain of the Urinary Constituents.

The estimates given in this part of the table are roughly ap- proximative, and represent the widest variations consistent with

normal conditions. The variations, always considerable, are
particularly marked as regards the uric acid.

```
Urea ...............................  ............................... 6·50 to 10·50 grains.
Chlorine (1·30 to 3·60 grs. of chloride of sodium)......... 0·80  "   2·15   "
Sulphuric acid (1·30 to 3·20 grs. of sulphates)  ......... ... 0·68  "   1·62   "
Phosphoric acid (2·10 to 4·00 grs. of phosphates).......... 1·17  "   2·25   "
    Do    do  combined with alkalies (phosphate of
                 soda and phosphate of magnesia).... 0·78  "   1·40   "
    Do    do  combined with earths (phosphate of
                 lime and ammonio-magnesian phos-
                 phate).. ............................. 0·39  "   0·85   "
Uric acid (0.40 to 0·70 grs. of urates)..................... 0·23  "   0·40   "
```

TABLE FOR REDUCING THE INDICATIONS OF A GLASS URINOMETER TO
STANDARD TEMPERATURE (60° Fahr.) WHEN THE SPECIFIC GRAVITY HAS
BEEN TAKEN AT A HIGHER TEMPERATURE. (BIRD, *Urinary Deposits*,
etc., Philadelphia, 1859 p. 7J.)

Temperature and degree.	No. to be added to the indication.	Temperature and degree.	No. to be added to the indication.	Temperature and degree.	No. to be added to the indication.
60	0·00	69	0·80	78	1·70
61	0·08	70	0.90	79	1·80
62	0·16	71	1·00	80	1·90
63	0·24	72	1·10	81	2·00
64	0·32	73	1·20	82	2·10
95	0·40	74	1·30	83	2·20
66	0·50	75	1·40	84	2·30
67	0·60	76	1·50	85	2·40
68	0·70	77	1·60		

APPENDIX.

As a matter of convenience to the student we present the addresses of American dealers in Microscope Stands, Objectives, etc. The list is compiled from the various catalogues and other sources of information which happened to be in our possession. It is therefore probably incomplete, but nevertheless will be found to comprise most, if not all of the leading dealers. Of the latter, those marked with an asterisk (*) issue illustrated catalogues, which can be obtained on application, free of expense:

NAMES AND ADDRESS OF DEALERS IN MICROSCOPES, OBJECTIVES, ETC., ALPHABETICALLY ARRANGED.

(*)Bausch & Lomb Optical Company, 37 Maiden Lane, New York; factory at Rochester, New York.

(*)R. & J. Beck, (London); American Agency in charge of W. H. Walmsley, Chestnut street, Philadelphia, Pa.

(*)W. H. Bulloch, 126 Clark street, Chicago, Ill.

F. J. Emmerich (Importer), 38 Maiden Lane, New York.

E. Gundlach, L. R. Sexton, agent, Rochester, New York.

(*)T. H. McAllister, 49 Nassau street, New York.

———— Pike (Dealer), 518 Broadway, New York.

(*)James W. Queen & Co., 924 Chestnut street, Philadelphia, Pa.

William A. Rogers (Micrometer Plates, etc.,) Cambridge, Mass.

(*)Messrs. C. A. Spencer & Sons, (Objectives, Accessories, etc.,) Geneva New York.

L. Schrauer, 50 Chatham street, New York.

(*)J. W. Sidle & Co., Lancaster, Pa.

(*)Robert B. Tolles, Hanover street, Boston, Mass. Charles Stodder, Agent, Devonshire street, Boston, Mass.

(*)George Wale, Patterson, N. J.   George Wale & Co., Hoboken, N. J., Agents.

(*)William Wales, Fort Lee, New Jersey.

(*)Joseph Zentmayer, 147 South Fourth street, Philadelphia, Pa.

The following price list of stands, etc., is only intended to include those previously described in this book. Several of the makers therein mentioned manufacture various models not included in our list. The stands ot the Messrs. Spencer's have been intentionally omitted, as we learn that they are now devoting their entire attention to the production of objectives.

The list of objectives is similarly incomplete. With the glasses of Messrs. Tolles and Spencer we have had a large experience. Several months ago the Messrs. Bausch & Lomb sent us a line of their objectives for study, all of which, after working with the same more or less for nearly a year, proved reliable glasses for their cost.

The entire list may be understood to include only such objectives as we have used to a greater or less extent in practical work, and of which we can therefore speak from experience, and without prejudice to other reliable makers.

We earnestly advise the student contemplating the purchase of object glasses, especially when ordering the same to be made by the optician, to seek the advice of some expert friend. Protection is thus afforded both to the maker and the buyer, and it has often happened that the latter, through sheer incompetency, returns a really fine glass to its maker, with the assertion that it has proved unsatisfactory. While on the other hand, especially in glasses of high balsam angles, there IS a choice in the work of the very best makers. The expert, too, will render valuable assistance in specifying the exact kind of objective desired.

MICROSCOPE STANDS. BY W. H. BULLOCH, NO. 126 CLARK STREET, CHICAGO, ILL.

Large best stand, A1, binocular. iris diaphragm, with the latest improvements, draw-tube, 5 eye pieces.........	$300 00
Same as above, but monocular, iris diaphragm, with three eye pieces..	250 00
Polished mahogany case, with side case for accessories.	25 00
Small best stand, A. B., binocular, 4 eye pieces and iris diaphragm.......................................	225 00
Monocular, two eye pieces and iris diaphragm.........	175 00
Polished mahogony case............................	20 00
" D " stand. all brass, 2 eye pieces and case	75 00
" D " stand, all brass except the base, 2 eye pieces and case...	67 00

MICROSCOPES AND OBJECTIVES BY THE BAUSCH & LOMB OPTICAL COMPANY.

The Professional Microscope, with the following accessories : Hemispherical immersión condenser and oblique light projector, plain and concave mirrors, sub-stage, receiving accessories of standard English size, two revolving diaphragms, sub-stage with internal " society screw," 2 slot diaphragms for mirror, 3 Gundlach's periscopic eye pieces (B, C and D), four objectives, viz : 2-inch, 1-5th inch, 3-4th inch and 1-8th inch immersion, with adjustment for cover, magnifying powers from 30 to 800 diameters, eye piece micrometer, camera lucida, bull's eye condenser, the whole in upright walnut case, with handle, lock and key, and drawer for accessories....................... $200 00

Large Student's Microscope, plain and concave mirrors, sub-stage of extra size to receive standards, English accessories, revolving diaphragm, etc., all attached to the swinging mirror bar ; sub-stage and mirror are removable ; three eye pieces, A, B and C, three objectives, 2-inch, 3-4th inch and 1-5th inch, magifying from 22 to 375 diameters, eye piece, micrometer, camera lucida, in upright walnut case, with handle, lock and key, and drawer for accessories 90 00

Family Microscope, concave mirror, adjustable for oblique light, one (B) eye piece, one first-class achromatic objective (one-half inch dividing) powers, 50 and 100 diameters, in upright walnut case, with handle, lock and key 20 00

The Messrs. Bausch & Lomb have brought out a new model which they term the Physicians' Stand, which may be thus described :

Large, heavy cast iron foot, rack and pinion for coarse adjustment, draw-tube, allowing 2 1-2-inches shortening, fine adjustment by a new frictionless motion, large hard rubber stage, sub-stage of standard size, carrying

three diaphragms, which, when pushed up, will closely
reach the object slide, plain and concave mirrors, arranged so that their distance from the object may be
varied, 2 eye pieces, A. and C, 2 objectives, 3-4ths inch
and 1-5th inch, magnifying powers, when the tubes are
completely drawn out, from 50 to 375 diameters, eye
piece, micrometer and camera lucida, in upright walnut case, with lock and key.......................... 60 00

The past twelve-month has witnessed unusual activity on the
part of microscope makers. Mr. Zentmayer, in the production
of his " Histological," (as also the advent of the Acme Stand)
stimulated other makers to increased energy in the production
of small, low-priced and *reliable* stands capable of performing
the work of larger and costly instruments.

At the late (August, 1880) meeting of the American Society
of Microscopists, held at Detroit, Michigan, two medium-priced
stands were exhibited which were regarded by competent
judges as being quite equal to the Acme Stand. One of these
was made by Mr. Bullock, of Chicago, and its cost—as near as
I can learn— will be about $50.

The other stand referred to was exhibited by the Messrs.
Bausch & Lomb, of Rochester, N. Y. It is called the " Investigator." Messrs. B. & L. have furnished the author with the
following description of

THE INVESTIGATOR MICROSCOPE.

In this stand we confidently claim to have reached a higher
degree of perfection than is possessed by any one approximat-
ing it in price. We have combined in it the qualities of a
first-class and high-priced stand, and at no sacrifice of its
working qualities. The different parts are ingeniously combined, are strong and firm, and in the parts subject to friction
we have introduced, as much as possible, new compensating
bearings, which insure the retention of smooth and reliable
movements with any amount of work. Working microscopists
will understand the value of this quality. When contracted it
stands but 11 high, but can be extended to 18 inches.

The base is of the tripod form, neatly japanned ; pillar and arm of brass, connected by a solid joint, which allows inclination of body to any angle; rack and pinion for coarse adjustment, fine adjustment by our patent frictionless motion ; main tube with two draw tubes. This is an entirely new feature in microscopes, which is an unquestionable improvement. It permits the use of standard length of tube for quick adjustment in outside tube, same as in instruments without rack and pinion adjustment; the same for any low power objective and the use of amplifier in either combination. The outside tube has a broad gauged screw, and adapter with society screw. The stage lies in the same plane as center of movement for mirror, is of brass, and has concentric, revolving motion with removable clips. It is thin to allow great obliquty, and as it rests upon a strong projecting arm, is perfectly firm under any manipulation. The mirror bar swings with a perfectly easy but firm motion, upon one bearing to any obliquity below, and above the stage for the illumination of opaque objects, and has affixed to it a secondary bar, to which the mirror is attached and which allows the separate use of the latter in any position of the sub-stage. It is provided with a sliding arrangement, whereby the mirror may be moved to and from the object. The mirrors are plain and concave, and of large size. The substage is adjustable along the mirror bar, and entire removable. It contains a diaphragm which may be brought directly under the stage. The ring is of standard size and is easily centered by a set screw. The stand is furnished with an eye piece of any power.

In black walnut case, with receptacles for eye pieces and objectives, drawer for accessories, handle and lock, price $40.00.

Same, with C Eye Piece, arranged for micrometer, camera lucida, micrometer, 3-4 inch and 1-5 inch objectives, $65.00.

BAUSCH & LOMB OBJECTIVES.

	Deg.	Price.
2 inch	18	$12 00
2 inch	12	6 00
1 inch	36	20 00
1 inch	20	6 00

	Deg.		Price.
3-4 inch	27	....	8 00
1-2 inch	72	....	22 00
1-2 inch	40	....	9 00
4-10	100	....	20 00
3-10	75	....	11 00
1-4	100	....	14 00
1-5	108	....	15 00
1-6 inch	180	Adjusting	30 00
1-8 Immersion.............	180	Adjusting	40 00

MICROSCOPES OF THE MESSRS. R. & J. BECK, OF LONDON. AMERICAN AGENCY IN CHARGE OF W. H. WALMSLEY, ESQ., 1016 CHESTNUT STREET, PHILADELPHIA, PA.

Popular Microscope stand, binocular, with one pair of
eye pieces, concave mirror, diaphragm, forceps, glass
plate, pliers, etc... .... $65 00

Same, but monocular 40 00

Economic Microscope, as furnished by the makers, monocular, with sliding coarse adjustment, 1 inch and
1-4 inch object glasses, 1 eye piece, concave mirror,
condensing lens, glass plate with ledge, brass pliers
and diaphragm, in mahogany case................... 35 00

The same, with rack and pinion, for coarse adjustment,
concave and plane mirrors, stage forceps, etc., in mahogany case.................................... 50 00

(Both of the above stands are fitted with fine adjustment screw.)

Extra eye pieces, each 4 00

The Messrs. Beck have just published a large and complete illustrated catalogue of their microscopes and
other scientific instruments, which they furnish to any
one on application. Among their later models of microscope stands we notice the "New National," costing, with one eye piece, concave and plane mirrors,
diaphragm, stage forceps, glass plate pliers, etc 40 00

Mahogany cabinet for the same...................... 10 00

Since the chapter on stands was written we have received the following description of

BECK'S INTERNATIONAL MICROSCOPE STAND.

The improved large best or international microscope stand has a tripod (A) for its base, upon which is placed a revolving fitting (B), graduated to degrees, by which means the microscope can be turned round without its being lifted from the table, and the amount of such rotation registered; upon this fitting two pillars are firmly fixed, and between them the limb (C) can be elevated or depressed to any angle, and tightened in its position by the lever (D). The limb carries at one end the body (E) (binocular or monocular) with eye-pieces and object-glasses; in its centre the compound stage (F), beneath which is the circular plate, sliding on a dove-tailed fitting, and moved up and down by the lever (Z), and carrying the supplementary body or sub-stage (G); and at the lower end a triangular bar carrying the mirror (H). Each of these parts requires a separate description.

The binocular body consists of two tubes, the one fitted in the optical axis of the microscope, and the other oblique. At their lower end and immediately above the object-glass there is an opening, into which a small brass box or fitting (I) slides; this box holds a prism so constructed that when slid in it intercepts half the rays from the object-glass, diverts them from their direct course, and reflects them into the additonal or oblique tube. To the prism-box is attached a spring-catch, which, when pressed in, permits of the removal of the prism-box; but this is only needed for cleaning, as, when the box is drawn back to the distance allowed by this spring, the prism in no way interferes with the field of view, and all the rays pass up the direct body, and the microscope is converted into a monocular one.

The upper or eye-piece ends of the tubes are fitted with racks and pinion for varying the distances between the two eye-pieces to suit the differences between the eyes of various persons; and arrangements are made for racking out one tube more than the other, to suit irregularities or inequalities between the eyes of the observer.

This body is moved up and down with a quick movement by

means of the milled heads (K), and with a very delicate and fine adjustment by the milled head (L). This milled head works against a lever, which moves a slide independent of the rack-movement, and gives an adjustment at once certain and decided.

The compound stage is of an entirely new construction: the object is most frequently merely placed upon it, but, if necessary, it can be clamped by carefully bringing down the spring-piece (M), the ledge will slide up or down, and the object may be pushed sideways; this arrangement forms the coarse adjustment. Finer movements in vertical and horizontal directions are effected by means of two milled heads (N and O), the screws attached to which are kept up to their work by opposing springs so as to avoid all strain or loss of time. The whole stage revolves in a circular ring by the milled head (P), or this can be drawn out, and then it turns rapidly by merely applying the fingers to the two ivory studs (Q Q) fastened to the top plate, which is divided into degrees to register the amount of revolution. The stage is attached to the limb on a pivot, and can be rotated to any angle, which angle is recorded on the divided plate (R), or can be turned completely over, so that the object can be viewed by light of any obliquity without any interference from the thickness of the stage.

Beneath and attached to the stage is an iris diaphragm (S), which can be altogether removed, as shown in the illustration, from its dove-tailed fitting, so as not to interfere during the rotation of the stage. The variations in the aperture of this diaphragm are made by a pinion working into a racked arc and adjusted by the milled head (T).

Beneath the stage are two triangular bars (U V), the one revolving round and the other rigid in the optical axis of the instrument. On the former the sub-stage (G), carrying all the apparatus hereafter described for illumination and polarization, fits, and is racked up and down by the milled head (W); the mirror also, if desired, slides on the same bar; the revolving motion to this bar is given by the milled head (X), and the amount of angular movement is recorded on the circle (Y), whilst the whole of this part of the stand is raised and lowered concentric with the optical axis of the instrument by the lever (Z), and the amount of such elevation or depression is registered on a scale attached to the limb. This bar can be carried round and above the stage, and be thus used for opaque illumination.

The lower triangle bar (V) carries the mirror (H), or a right-angle prism, when the illumination is required to be concentric with the optical axis of the instrument, and independent of the movements of other illuminating aparatus.

The mirror-box contains two mirrors, one flat and the other concave; it swings in a rotating semi-circle attached to a lengthening bar, which enables it to be turned from one side to the other, and revolves on a circular fitting for giving greater facilities in regulating the direction of the beam of light reflected, the whole sliding upon either of the triangle bars, previously referred to, and made to reverse in the socket (a) so as to bring the centre of the mirror concentric with the axis of the microscope in either case.

As the mirror alone is insufficient for many kinds of illumination, some provision has to be made for holding various pieces of apparatus between the object and the mirror. For this purpose a supplementary body, or sub-stage, is mounted perfectly true with the body, and is moved up and down in its fitting by rack and pinion connected with the milled heads (W). This sub-stage, to which reference has already been made, is now regarded as one of the most important parts of the achromatic microscope; in it all the varied appliances for modifying the character and direction of the light are fitted. But a few years since it was considered sufficient for this part of the stand to be constructed so as to move up and down perfectly coincident with the optical axis of the instrument, and for that purpose it was racked in a groove planed out on the same limb as that on the upper end of which the optical portions were carried. But lately microscopists have shown the desirability of affording every facility for lateral angular adjustments; and this has led to the sub-stage being attached to an arc (b) working in the circular plate (Y), and moved by a rack and pinion (X), whilst the amount of such angular movement is recorded on the upper surface of the plate (Y). Having once fixed the angular direction of the light, the focussing of it depends upon the lever (Z), which moves the circle up and down, and with it the arm carry- ing the illuminating apparatus *in the optical axis of the instru- ment.* So long ago as 1851 Mr. Grubb, of Dublin, called atten- tion to the advantage of mounting the illuminating apparatus on a revolving arm or arc, which he thus describes in his pro- visional specification for improvements in microscopes. No. 1477, 5th July, 1851 :—" My third improvement consists in the addition of a graduated sectorial arc to microscope concentric to the plane of the object "*in situ,*" on which either the afore- said prism or other suitable illuminator is made to slide, thereby producing every kind of illumination required for microscopic examination, and also the means of registering or applying any definite angle of illumination at pleasure." With but slight modification, this is the plan adopted in this stand.

The sub-stage is also fitted with complete centering and rotat- ing adjustments, the latter having a graduated circle attached, and fittings for carrying Darker's series of selenites, blue glass disks for modifying the light, etc. In all the requirements of an instrument of precision, and fully meeting the wants of the most advanced modern workers, it is confidently believed that this new stand has no rival.

The price of the *International Stand, Binocular,* with five eye- pieces, concave and plane mirrors, and right angle prism, stage forceps, pliers and glass plate with ledge, is . . $325.00

The price of the *International Stand, Monocular,* with three eye-pieces, and the same fittings as above, is, . . $275.00

(The Messrs. Beck manufacture several grades of objectives, as will be seen by reference to their published catalogue.)

J. W. SIDLE & CO., LANCASTER, PA., ACME MICRO- SCOPES AND ACCESSORIES.

We here quote the prices of the Acmes, and a few of the more important accessories.

3*a.* Acme Monocular, with brass tripod base, rotating stage, iris diaphragm and substage fitting with Society screw for using objectives as condensers ; in walnut case with lock and handle, and drawer for accessories, $ 50.00

3b. Acme Binocular, with one pair oculars, and outfit as
above, ...   75.00
3c. Acme Monocular, as above, with ¼ objective of 32°
and ¼ of 100°, and small bull's eye condenser,   .   .   75.00
3d. Acme Binocular, as above, with ¼ of 32°, ¼ of 100°
and bull's eye, ...   100.00
2a. Large Acme Monocular, rotating glass stage, sub-
stage, with rack and pinion, in polished mahogany case
with side case for accessories, ...   100.00
2b. Large Acme Binocular, same as above, with additional nose-piece, as per circular, ...   135.00
2c. Large Acme Monocular, as above, with mechanical
stage and 2 pair oculars, ...   135.00
2d. Large Acme Binocular, as above, with mechanical
stage and 2 pair oculars, ...   170.00

PRICES OF ACCESSORIES.

Mechanical stage for Acme No. 3, ... $16.00
Walnut base with lamp and adjustable stand, with brass
fitting to receive lower end of pillar, converting the Ac-
me into a Class Microscope, ...   5.50
Camera Lucida, ...   5.50
Neutral Tint (Beale's), ...   3.00
Iris Diaphragm, ...   4.50
Sub-stage fitting for using objectives as condensers,   .   1.00
Woodward's prism, unmounted, ...   1.50
       "   "   mounted to screw in stage,   .   .   4 00
Mechanical Stage, ...   25.00
Sliding Object Carrier. fitted to stand, ...   4.00
Paraboloid, plain, ...   8.50
Parabolic Illuminator, ...   8.50
Polarizer for Acme, selenite screwing in stage, to rotate
independently of Nichol prism, ...   13.50
Polarizer, best, with large prism, ...   18.00
Mechanical Finger, ...   6.50
Solid Ocular, ¼, ⅓, ½ inch, ...   8.00

CONGRESS TURN TABLE.

This table has made a reputation for itself. Retail price... $6.50

PRICES OF ACME SERIES OF OBJECTIVES.

2 inch,	15° . . . $13.00	3-10 inch, 115° non adjustable, . . $16.00
1½ "	18° . . . 14.00	3-10 " 115° adjustable, 20.00
1 "	36° . . . 15.00	1-4 " 100° non-adjustable, . . 14.00
⅝ "	82° . . . 12.00	1-5 " 120° non-adjustable, . . 16.00
⅜ "	36° . . . 15.00	1-5 " 120° adjustable, 20.00
¼ "	60° . . . 14.00	Other lenses quoted in Catalogue.

MICROSCOPE OBJECTIVES, BY MESSRS. C. A. SPENCER & SONS, GENEVA, N. Y.

FIRST CLASS.

Size.	Deg.	Price.
3 inch	13	$18 00
2 inch	20	25 00
1 inch	50	45 00
2-3 inch	47	32 00
1-2 inch	100 Adjustable	50 00
1-4 inch, dry and immersion	180	65 00
1-6 inch, " "	180	70 00
1-8 inch, " "	180	70 00
1-10 inch, " "	180	80 00
1-16 inch, " "	180	115 00

PROFESSIONAL SERIES.

3 inch	13	$18 00
2 inch	16	18 00
1 inch	33	18 00
2-3 inch	36	20 00

23 Microscopy.

1-2 inch	65	25 00
1-4 inch, Adjustable.......	115	24 00
1-6 inch, "	175 Dry and immersion.....	34 00
1-8 inch, "	175 " "	40 00
1-15 inch, "	175 " "	50 00

STUDENTS' SERIES.

3 inch....................	8	$6 00
2 inch....................	10	6 00
1 inch....................	22	10 00
2-3 inch	32	12 00
1-2 inch	50	15 00
1-4 inch	100	16 00
1-8 inch	135	25 00
Wenham's Reflex Illuminator........................		10 00

(The Messrs. Spencer & Sons furnish a full line of goods desirable by the microscopist, such as solid eye pieces camera lucidas, and accessories generally. Their former connection with the Optical Company has been dissolved, and they now continue business on their own account, as above stated, at Geneva, N.Y.)

MICROSCOPES AND OBJECTIVES, BY R. B. TOLLES, BOSTON, MASS.

Tolles' Large Microscope, "B," no objectives	$225 00
Student's stand, 1 inch E. P., 1, and 1-4th inch objectives, packed in black walnut case....................	50 00

Additions, etc.. to the above—

Extra eye pieces, each........................	4 00
Camera lucida	5 00
Sub-stage, for accessory apparatus......................	5 00
Sliding stage, giving horizontal and vertical movements by hand..	15 00
Fine adjustment by lever and micrometer screw........	20 00
Rack and pinion for coarse adjustment.................	12 00
Draw tube	4 00
Plain mirror ....·····..............................	3 00
Thin glass stage to rotate on the optical axis	10 00
Packing boxes for transportation, extra	1 00

Student's Microscope, with one inch and one-fourth
inch objectives (second quality) one ocular, rack and
pinion, lever, fine adjustment for focus, sub-stage for
Illuminating apparatus, revolving diaphragm, plain
and concave mirrors, side stand for illuminating
opaque objects, black walnut case 90 00

TOLLES' FIRST CLASS OBJECTIVES.

Size	Degrees.	Price.
1-2 inch, Angle apertured....	60 to 90	$40 00
4-10 " "	90 to 120	45 00
4-10 " "	135 to 145	65 00
1-4 or 1-5 " "	120 to 130	45 00
1-4 " "	150	55 00
1-4 " "	180	70 00
1-6 " "	180	75 00
1-8 " "	180	80 00
1-10 " "	180	85 00

TOLLES' PATENT SOLID EYE PIECES.

1-2 inch, " D "............................ .... $ 8 00
1-4 inch.. 8 00
Either of the above, for his Student's Microscope 6 00
Wenham's Reflex Illuminator 10 00

MICROSCOPE STANDS, BY JOSEPH ZENTMAYER, 147 SOUTH FOURTH STREET, PHILADELPHIA, PA.

American Centennial Stand, binocular, with 5 eye pieces $300 00
Same, but monocular, 3 eye pieces..................... 250 00
Concentric adjustable stage, extra..................... 20 00
Best mahogany case, with fine handle and side case 30 00
American Histological Stand, with plain and concave
mirrors attached to swinging radial bar, carrying, also,
sub-stage and diaphragms having three apertures, 1 (B)
eye piece, no case 30 00
Extra eye pieces, each 5 00

NOTE.—The following extras can be obtained of Mr.
Zentmayer at the prices below given. They are fur-

nished at the suggestion of the author, who invariably orders them for his friends and pupils. The instrument is thus rendered efficient for all the ordinary purposes of the working microscopist :

Supplemental revolving stage........................... $1 00
Sub-stage adapter, to carry condenser, society screw.... 1 00
Eye-piece adapters, ready to carry solid cells, each...... 1 00

STAGE MICROMETERS, TEST PLATES, ETC., BY W. A. ROGERS, CAMBRIDGE MASS.

Mr. Rogers has just completed his new machine, in the construction of which several important improvements have been secured. We cheerfully add our endorsement as to the regularity and delicacy of his rulings. One of his plates, ruled up to 120,000 lines to the inch, now in our possession, is a marvel of art. Owing to the extreme shallowness of Mr. Rogers' finer rulings, they are more difficult tests than similar bands by Nobert. Mr. Rogers furnishes a great variety of work. Some of his regular plates may be thus described :

Stage Micrometer, consisting of 10 lines, 100, and 10
   lines, 1,000 to the inch $1 25
Stage Micrometer, consisting of 25 lines, 200 to the inch,
   10 lines, 1,000, 10 lines, 2,000, 10 lines, 4,000, 10 lines,
   5,000, 10 lines, 10,000, 10 lines, 20,000, and 10 lines, 30,000
   to the inch ....:............................... 4 00
Stage Micrometers, ruled 1,000, 10,000, 20,000, 30,000, 40,000
   and 50,000 to the inch 5 00
Same as above, to 60,000 .... 6 00
    "   "   to 70,000, $7.00; to 80,000, $8.00; to 90,000   9 00
    "   "   to 100,000 .... 10 00
   (Closer rulings by special contract.)
Standard half-inches, with 50 subdivisions of 1-100th inch
   and 10 subdivisions in the centre of 1-100th inch ....: 3 00
Standard centimeters, with 100 subdivisions 4 00

(At double the above prices Mr. Rogers furnishes a complete discussion of the errors of the plates, and a table showing the corrections to be applied to each ruled space to reduce it to the United States standard.)

Standard half metres and standard metres, either on glass, steel, iron or nickel surface, from $25.00 to...... 100 00

Test plates, micrometers, etc., are also ruled by Mr. Charles Fasoldt, 594 Broadway, Troy, N.Y. The author has one of Mr Fasoldt's 120,000 band plates, which he values very highly. He regrets not being able to give Mr. Fasoldt's Catalogue of Prices.

SUPPLEMENT.

CONTRIBUTIONS TO THE CINCINNATI MEDICAL NEWS.

Inasmuch as the author's contributions during the past six years to the pages of the *Cincinnati Medical News*, and also the *American Journal of Microscopy* contain instructions more or less interesting to those commencing the study of the microscope, it is thought desirable to present them in this supplement. It was however found on inquiry that the back numbers of the Cincinnati journal were not to be obtained.

Dr. Blackham, of Dunkirk, N. Y., kindly placed his bound volumes of the *News* in the hands of the writer, thus enabling him to give the reader the following references, dates, etc., to the several articles contained in the *Medical News*.

1875. January, "Microscope Objectives," p. 41; February "Measurements of Moller Probbe Plate," p. 92; March "Wide *vs.* Low Angled Objectives," p. 129.; April "Diagnosis of Blood Stains, Dr. Woodward *vs.* Dr. Richardson," p. 177.; April, "A Simple Method of Demonstrating Ang. Aperture," p. 183.; May, "Wide *vs.* Low Angled Objectives," p. 228.; June, "Dr. Chester Morris' Reports on Objectives," p. 281.; November, "Wide *vs.* Low Angled Objectives," p. 515.

1876. January, "Battle of the Object-Glasses," p. 41.; February, "The Mic. and Its Mis-interpretations," p. 86.; March, "A New Microscope Stage," p. 232.; April, "Nobert's 19th Band," p. 237.; July, "Wythe's Amplifiers," p. 433.; August, "Prof. J. E. Smith *vs.* the Nachet 1-5th," p. 485.; August, "Trichiniasis," p. 488. October, "Necessary Manipulation of the Microscope; to Exhibit Fine Striæ," p. 533.; December, "Torula Cerevisæ in Human Urine," p. 677.

1877. January, " Wenham's Reflex Illuminator," p. 74.; July, " R. Hitchcock, Esq., *vs.* High Angles," p. 504.

The past contributions to the *American Journal of Microscopy* will, owing to the kindness of Prof. Phin be given in full.

PROF. R. HITCHCOCK VS. HIGH ANGLES.

In the May number of the *American Journal of Microscopy*, I find an article containing a good-natured criticism of a paper read by me before the Dunkirk Microscopical Society last October.

Mr. Hitchcock candidly states that he has only seen a short abstract of this paper, and has but an imperfect knowledge of it. He further suggests that his main object was to call further attention to my views. and he suggests that I put them in form, so as to be published in the *American Journal*.

For the benefit of Mr. Hitchcock, I will state. that my views have been clearly stated in a series of articles, which have been published in the Cincinnati *Medical News* during the past three years, under the caption of " High *vs.* Low Angles," and have thence been quoted from and reprinted in various publications ; certainly I cannot be expected to go over the ground anew at this late day. It is true that I exhibited before the Dunkirk Microscopical Society the No. 20 of the balsam Moller probe platte, and also the 19th Nobert band, both so plainly that all who were present saw without difficulty. These tests were *not* difficult for the glasses employed, as was attested by the fact that they were shown in a crowded room, amid the attendant jars and vibrations, my object being to *demonstrate* the facility with which my Tolles' duplex glasses handled these so-called difficult tests.

To show the work of these same lenses by central light, I selected the same test, *Nav. Angulatum* which the biological committee at Philadelphia has declared *impossible* to be shown at a less angle than 20 degrees from axis, with any medium power glass. This test was displayed illuminated by light through a central aperture placed close to the object, just large enough to light the field, the diameter of the aperture being about 1-200 of an inch, direct light being used, *i. e.*, without mirror or condenser. The ease with which the duplex handled this test was made amusing and apparent by my picking up the stand, walking around the room, sitting it down again haphazard before the lamp, when the resolution was found by a gentleman appointed to examine to be unimpaired. This was repeated three times. Will Mr. Hitchcock repeat, using a low angle glass, be it a one-fifth or a one-fiftieth, and report?

I had other tests for central light work, including histological and pathological preparations. It was impossible to show all of these to so many people as were present, as it was, the entire evening after the reading of my paper was occupied. Hence, it will be seen that I fought the low angles on their own *chosen* ground, and with the express view of demonstrating that the very best preformance of the duplex lenses is seen by central light,

· Mr. Hitchcock further says that " the universal testimony of our best authorities, who have spent their lives in microscopical work, is against Prof Smith."

It was just this kind of testimony that affirmed a few years ago, the highest *possible* aperture of an object glass to be 135 degrees; that the resolution of the Nobert 19th band was a matter of faith rather than of sight, etc., etc. Mr. Hitchcock is welcome to the witnesses.

Note this fact, to-wit, in original investigations, the advanced worker must necessarily be in a minority. I rejoice that some of Mr. Hitchcock's witnesses have lately found cause to change their opinions. Dr. Carpenter will no longer assert that the resolutions of the Nobert 19th band is a " matter of faith, rather than of sight." On the other hand, he has given unqualified endorsement to the superiority of the duplex glasses.

Mr. Hitchcock desires to ask *why* I think that most of the work in histology and pathology already " done" with the so-called " working lenses" of narrow angles would require further attention, and with wide angled glasses I reply, that in the past four years great advances have been made in the construction of objectives, and in the manipulation of the microscope; what was considered a " good working glass" ten years ago, would *now* be totally valueless for advanced work. We now demand, as near as may be, perfect lenses, and superior dexterity in handling them; and these two conditions are inseparable. The finer the objective the louder the call for expert manipulations. That well-known term, " working lenses of narrow angles" means, when stripped stark naked, easy going lenses, with no screw collar to bother; good working lenses, that a child or sleepy adult without experience can use right along, will work through covers of common window glass, big working distance, and all that, etc., etc. Such *are* admirably adapted to the use of those who use the microscope as a plaything; admirable things, too, to prove that " a little knowledge is a dangerous thing."

Be it known, that I do not condemn an objective simply because it has a narrow aperture; conversely, I do not endorse a glass on account of its wide angle. I have seen scores of wide angle glasses not worth the cost of their brass mountings. As to " errors in interpretation," the more perfect the lens, and the more expert the manipulator, the less chance of error. Under high amplifications, a superior wide-angled glass, *properly handled*, will generally prove the more reliable; and in advanced work *cannot* be dispensed with, be the illumination central or oblique.

Finally, I have to thank Mr. Hitchcock for his friendly criticism. He seems, evidently, to be after the facts. I have responded to his request as well as I could with my limited time and space. Two hours "over the tube" would demonstrate more than volumes of print.

J. EDWARDS SMITH.

Note.—On my last visit to the Dunkirk Society, May, 1877, I showed, I believe, for the first time, the Nobert 19th band as an *opaque object*, with my Tolles' one-tenth duplex, Beck's *vertical* illuminator being used. It is obvious that the pencils of light traversing that band were at least centrally disposed. J. E. S.

ANGULAR APERTURE AND CENTRAL ILLUMINATION.

In my reply to Prof. R. Hitchcock, reprinted in the August number of this journal, I endeavored to respond to his courteous request. My position had been placed fairly before the public for years, and I was, for the nonce, content in maintaining silence, feeling assured that when attention should be thoroughly aroused as to the claims of high-angled glasses, that I would be better understood. I think I can to-day safely assert that my expectations have been realized, and to a greater extent than I had any reason to anticipate.

Prof. Hitchcock has done me the honor of an extended reply in the September number of this journal. It is noticeable that he still preserves the same courteous bearing which characterized his previous paper; indeed he pays me a compliment ("We have his plain statements, and we accept," etc., p. 110). All this, as an index of his good nature, is very acceptable.

Nevertheless, be it remembered that it has been no part of my purpose to obtain acceptance *per se* of my positions. On the contrary it has been my earnest endeavor to excite attention, and to induce my readers to experiment for themselves. Hence the results of my tedious and protracted experience with the duplex lenses were printed in forcible terms (heterodox as they must have seemed to many), allowing no loop-hole for retraction or qualification on my part. Hence it was, too, that I requested Prof. Hitchcock to repeat some of my experiments; Prof. H. is compelled to decline, because he has not the objectives at command. This I sincerely regret.

Premising that my time is fully occupied, and that it will be impossible to reply directly to Prof. Hitchcock's last, and interesting paper, I proceed to offer a few thoughts suggested by the same.

First, what, in common parlance, constitutes a high-angled objective? I think I understand Prof. Hitchcock—perhaps not—let us see. It may be that in arriving at a mutual understanding on this point, we may get, generally, nearer coincidence. Now, most observers associate with the term " high-angled objectives," some great display of figures. Thus, 175°, 178°, or the " impossible " 180°. Others might go still further, turning the 180° corner, and thus revel in the balsam angles, say, up to 100° or higher. Such are not exactly my own notions.

I regard as high-angled, all objectives from a 4-inch to a

1-75th, having respectively the highest possible attainable aperture. If this definition be accepted, then it occurs that what was known as a high-angled 1-10th of 130° ten years ago, would now be classed as a medium-angled glass, and further, that an inch objective of 45° would take rank as a high-angled glass, and to the latter class of objectives do I refer as giving me the best results for any and all work, selecting from this class of objectives the one best adapted for the work in hand. Let me illustrate by selecting a case that I almost daily have in practice —to-wit:

Suppose, for instance, I am making preparations, say, of malignant growths (notice that this is not a diatom), it would doubtless be desirable first to make a preliminary examination. I should, from the high-angled class of objectives, select the above named inch of 45°. Why?

First, because it has large working distance, which is not disturbed by high eye-piecing. I am also enabled to bring the mirror above the stage, and thus easily condense light on the object; and further, the working distance is so large that little changes in the same do not disturb the object to such an objectionable extent as would occur had a higher power been selected.

Second, because this inch will show me structure up to 30,-000 lines to the inch. It will not only bear the 1-4th solid eyepiece, but will give with it added force of definition, with increased amplification.

Third, (result of first and second), because of all glasses I have ever handled, I can get the most work (I have in hand) out of it. For instance, its large reach, due to its working distance (the so-called penetration), enables me to focus through the different planes of the object, while the entire specimen remains fairly in view. I am further enabled to search through my mount, with the least danger of allowing important structural elements to escape my attention—an important item.

Now, should such preliminary examination suggest the presence of structure calling for an objective capable of recognizing lines 100.000 to the inch; the inch of 45° would be removed, and a 1-6th, having the widest aperture known, and working distance of one-fiftieth of an inch, substituted. With this superior defining glass I have still working distance enough to illuminate my object by condensing sunlight on top of the cover, thus securing exquisite definition under amplification of more than 5,000 diameters. Here, again, with said 1-6th, I get better work than I can with any other style of glass.

Again, should it be desirable to cross-question my specimen still closer, it would become necessary to remove the 1-6th and substitute a bran-new 1-10th not yet a month old, the masterpiece of a young American optician. This 1-10th has also the widest aperture known, combining still more exquisite definition with sufficient working distance to allow the use of sunlight as above named, while its glorious performances by central light eclipse, yea, distance the work of any low-angled glass extant, this beyond the ghost of a doubt. (And right here I beg Prof. Hitchcock to be lenient, and to allow me a little margin; it's hard to keep cool over the central-light performance of this new 1-10th). But, with the employment of this 1-10th, we get into tolerably close quarters, the " difficulty of working " is enhanced. It won't do to put the screw-collar somewhere between " zero " and somewhere else. And then, too. there has got to be a good deal of that " handling " attended to that Prof. H. says will do so well to talk about before an audience. Prof. Hitchcock here applies the great American question, and asks if all this bother will pay. I reply, " Yea, verily !"

I shall be very grateful if Prof. Hitchcock will point me to glasses that will take the place of those above mentioned. It has so happened that those who have thus far undertaken this task (and some have traveled hundreds of miles to accomplish it) have resigned their previous ideas in my favor. In fact, I have scores of friends made in just this kind of a way, and I hope to add Prof. H. to the list.

Having thus given my ideas as to the nature and performance of high-angled objectives, I beg to add that I see no necessity for backing down from my original statements, although challenged by Prof. H. On the contrary, the advent of the new 1-10th serves to clinch the nail that I had previously well-driven home with the duplex. And I now place again on record my deep-seated convictions that the use of such lenses as I have described will be of the greatest use in histological observations.

Prof. Hitchcock's remarkable statement that " much of the very best class of work is being done with second-class lenses," is, I fear, a slip of the pen. I, for one, certainly cannot agree with him, even if so be, that I am found in a " minority."

Prof. Hitchcock has, I fear, found out my weak side, and punches me sorely with Helmholtz's formulas. He has got things nicely tabulated, to show at a glance just what can be done and what can't. Now, I own up that of all the annoying things on this mundane sphere (to me), these confounded tables take the lead. Years after others as well as myself, saw the 19th Nobert band as clearly as we saw anything, came the tables to prove that the thing was an utter impossibility! It was noticeable too that, year after year, the tables got changed, in fact, improved, to approximate more closely to the facts. *Verb. sap.*

Again, Mr. Hitchcock informs us that " to attain the best possible results from an objective, the angle of the illuminating pencil must be the same as the angular aperture of the objective." This, he says, is the " chosen ground " for any lens.

To put it mildly, this is another most remarkable statement. I dismiss it with the remark that my own experience teaches me that such manipulation would defeat ninety low angled glasses out of the hundred. He also, referring to my exhibition of the 19th Nobert band, with the vertical illuminator, suggests the employment of a diaphragm to the back lens, predicting that the resolution would " doubtless be greatly improved." I had already tried the experiment (and scores of others); the result was *nil.*

Prof. Hitchcock further says that " Prof. Smith has not taught us anything new about objectives." Well, doubtless a little setting back of this kind will be of benefit to me, and may prevent the growth of undue conceit. I confess, that I really did entertain the notion that some of my " shows " before the Dunkirk Microscopical Society *were* novel. For instance, the exhibition of the Nobert 19th band as an opaque object; ditto. the showing of 30,000 lines to the inch with an inch objective, and under an amplification of 740 diameters, etc.

It was my intention to pass over one or two points presented by Prof. Hitchcock, but fearing that silence may be misconstrued, a word or two may not be amiss. Prof. Hitchcock does not deny that Dr. Carpenter has *lately* (note that point) given his unqualified endorsement to the superiority of the duplex glasses, but he adds, "so does everybody else." This, too, must be another slip of the pen, for it is all, yea, more than I have claimed, for at the date of my Dunkirk lecture, the claims of the duplex were *not* admitted by everybody. During the past four years I have, myself, received hundreds of letters from as many sceptics.

Prof. Hitchcock states that he was taken "a little by surprise" by my claim that the "very best work of the duplex is seen by central light," and with singular tact allows the little *fact* to remain intact. This is too bad, for right here I expected the thickest of the fight. It was, in fact, "my chosen ground." Prof. H. is too much for me on tactics.

In this connection, I beg the reader to bear in mind that while Prof. H. admitted the superiority of wide angles for the resolution of "diatoms and Nobert's test-plates," he nevertheless held that this did not support my view that high angles are "universally preferable," (see this journal, May, 1877). To meet this point, I stated in my response (this journal, August, 1877,) that I had not omitted such tests as are generally considered "chosen ground" for small apertures, *i. e.*, what is generally understood and accepted as dead central illumination; and further, that among other tests I selected *Navicula angulata* illuminated by central light, because such resolution had been publicly declared impossible. In reply Prof. H. kindly admits, and in complimentary terms, all that I had claimed as to central light work, and he frankly adds, "they are just what we would naturally expect *a priori*." Having made this full admission, Prof. H. seems to repent a little, and reverts to the subject again (page 112), still yielding that the high angles are superior, both with oblique or central illumination, *but for resolutions only*, because he says, the low angles have the greater penetration. The nature of this, his residual error, is manifestly indicated in the present paper. *Less* than "two hours over the tube" would convince Prof. Hitchcock that " the very best performance (penetration, depth of focus, reach, or what not included) is seen by central light. I again invite investigation. The passage from Frey was original with Dr. Carpenter ten years ago. I have no objection to urge.

A word, and let it bo with due respect to Dr. Carpenter. His advanced age and failing eyesight, together with his many duties and cares, forbid that he should continue to be the authority he has been in the past. Singularly enough, the doctrines advanced by Dr. Carpenter twelve years ago are still endorsed by many to-day. This operates as a brake to the wheel of progress; surely Dr. C. is not to blame. Mr. Hitchcock asks if Dr. C. has changed his opinion since the fifth edition of his work ; in reply, I can say that the latest information I have bearing on this point is this ; about eighteen months since a friend of mine was in London attending medical lectures. He sought Dr. Carpenter's advice as to objectives ; Dr. C. advised him to go home and purchase the duplex, and my friend did so purchase. Be all this as it may, while I entertain the highest respect and admiration for his medical and scientific attainments—his life of unceasing toil in his profession—I cannot, and for the reasons presented, attach importance to Dr. Carpenter's endorsement of the duplex glasses. In other words, I should prefer the testimony of American experts.

To resume, Mr. Hitchcock further says that " two hours over the tube " has demonstrated something, he don't say exactly what. At anyrate he volunteers his thanks, which I joyfully accept. I should have said that " two hours *jointly* over the tube " would demonstrate more than volumes of print. 1 hope yet to enjoy such a *tete-a-tete* with friend Hitchcock, and promise in advance to delight him with the new wide-angled 1-10th before referred to. The new glass, too, confirms in a truly practical manner the positions I have heretofore taken in print.

In conclusion, I have again to thank Prof. Hitchcock for his good natured review of my previous article, and hope that he will find nothing in this hastily written article objectionable.

A FEW WORDS CONCERNING THE PERFORMANCE OF OBJEC-TIVES.

Ed. Am. Jour. Microscopy—Referring to the letters of Mr. E. Gundlach, printed in the August issue of the Cincinnati *Medical News*, it was noticed that Mr. Gundlach bases his optical laws on the performance of his own objectives, or at least he appeals confidently to his objectives to sustain his positions, as to working distance, resolution and angular aperture.

Mr. Gundlach certainly has the right to state the principles which govern him in the construction of object-glasses, and it would be well if other makers would follow his example. But his readers must keep constantly in mind that while the general principles he sets forth may not, as a rule, differ materially from those actuating other opticans, each and every one has a *handling* characteristically his *own*, and that a necessary result of his will be recognized in the *characteristic* performance of various so-called first-class objectives; and further, that between the extremes comprised in the " handling" above named, a large margin obtains, sufficient to warrant the acceptance of one objective to the exclusion of another.

The potency of these considerations is enhanced when we remember that the *slightest* superiority in an objective is of the highest importance. Beale says (" How to Work with the Microscope," p. 285) " It is certain that the slightest advantage in defining power ought not to be underrated. * * * Improvement in the means of observation is of the utmost importance, and however slight, always leads to the discovery of new facts."

Again Dr. Beale writes (p. 283) " The best object-glasses will define clearly and accurately bodies which, from their transparency, are quite invisible under objectives only slightly inferior to the first."

It therefore becomes apparent that these little differences, perhaps due to " handling" on the part of the optician, while they are invaluable to the working microscopist, may also be referable to the genius of the optician—to a particular method of " handling" known optical laws.

By way of illustrating these "little differences," it may be mentioned that among first-class objectives of the same magnification, aperture and resolving power, greater and lesser working distance will be found, and right here is the solution of the dogma of (so-called) penetration, for, of the glasses above mentioned, that one having the greater working distance will be endowed with the greater penetration. This is the whole thing in a nutshell, as set forth by the writer years ago.

If, in the comparison of two objectives, alike in magnification, aperture and resolving power, one shall be found to have the longer working distance, then is its superiority demonstrated, and, be it remembered, that comparisons of objectives ought always to be made in this kind of a way, otherwise any attempt at comparison would be as futile as if we were to compare a turnip with an orange.

Again, *if it can* be shown that a medium power glass, say a 1-6th or a 1-10th *having the widest aperture known*, and also comparatively large working distance, say 1-50th of an inch, will more than do any work that can be done with objectives of exceedingly high power, and correspondingly *short* working distance (excess of aperture being barred by the above named conditions), it will be obvious that the former will be eminently the superior glass.

Now the *facts* warrant the assertion that just such one-sixth and one-tenth are made, and by American opticians. The writer has had, and still has in his possession just such glasses. His first contribution to the columns of the *News*. four years since, set forth their claims, and he has never had occasion to make either modification or retraction of his original statement. When such occasion shall occur, the readers of this journal will be promptly advised; until then I claim (especially for the one-sixth) unequaled intensity of definition, widest known aperture, combined with large working distance—in other words, without that sacrifice of "working distance" alluded to by Mr. Gundlach.

I embrace this, my first opportunity, to say that I have lately examined at length a new duplex one-tenth, the work of a young American optician. This new one-tenth is similar in

24 Microscopy.

aperture and has greater working distance than my own one-tenth ; its work over such tests as the Nobert and Moller (balsam) plates was quite equal to that of its formidable competitor, while the views it gave by central light were simply glorious, excelling the old one-tenth.   This, as a first attempt on the part of a young optician, is truly encouraging.   These new duplex glasses will receive my earnest attention, and I may have, in the future, something more to say about them.

Allow me to further state, that, being desirous of acquainting myself with the work of our own opticians, I wrote to the Rochester firm, asking the loan of objectives for examination and study.   They responded promptly, sending me by return mail a full line of Mr. Gundlach's lowest priced "Student's" work. They certainly had no time to select exceptional glasses.   My time has been so thoroughly occupied, I have not yet given them the attention they deserve, but as far as they have been examined I am much pleased with them—a three-fourth inch at $6, a one-half inch at $8, and a three-tenths at $11, are very fine indeed for their cost.

ANGULAR APERTURE ONCE MORE.

Ed. Am. Jour. Microscopy.— The discussion between Prof. Hitchcock and myself, which, during the last few months you have done us the honor to print, seems to have become somewhat of a rambling nature.   Prof. Hitchcock " changes base " so often, that it would bother a streak of lightning to catch— much less corner him.   Be this as it may—and perchance he might return the compliment, I can only say that I have endeavored to stick to the points at issue—the merits of high, as compared with low angles, I have attempted to set forth from a *practical* standpoint, being tolerably familiar with the use of either style of glass.

I have tried my best to have Prof. Hitchcock examine high-angled glasses for himself, and report results, but, as your readers are aware, from his own assertions, Prof. Hitchcock does not claim a *practical* knowledge of them, and as a matter of course he cannot see what that "handling "(of which it will do so well " to talk about before an audience ") amounts to. Of a piece with this comes his statements that such and such positions of my own are " not proved," and that before he accepts them he " must *know why*."

I submit: that the only possible method of convincing Prof. Hitchcock will be the practical one, in this instance, however, quite impracticable, to-wit : " two (or more) hours *conjointly* over the tube," or until he shall have done me the honor to adopt my suggestion, and experiment with the high angles for himself; in the latter case, obviously, he ought not to expect first-class results from the wide angles until at least he shall have learned that something is due to " handling." I repeat that until one or the other expedient is adopted, we might continue the present discussion until doomsday, and on my part without a vistige of any chance of convincing Prof. Hitchcock. I therefore offer the following brief remarks by way of closing a controversy which, in the nature of things cannot lead to any fixed results.

Prof. Hitchcock asks, " Why strain a one-inch to see 80,000 lines to an inch, with a deep eye-piece, when the half-inch would do the same work without being pressed so hard ?"

Well, let us strip things " stark naked," and find out " why." First, with the inch I get better definition ; second, longer working distance ; and, third, because the inch, with deep eye-piecing " is not pressed so hard as the half-inch—and this is my reply to Prof. Hitchcock. As a matter of course he will come back at me with the stereotyped assertion, " I must know why." He says, too, " Until his (my) experiments are published and subjected to public examination they are not to be accepted, etc." Well, by the laws. my experiments *have been* both published and subjected to public examination. But, unfortunately, Prof. H. wasn't there ! Note that.

Further, I now propose to put the inch and half-inch to a practical test. Thus, let Prof. Rodgers, of Cambridge, rule a band of 40,000 lines to the inch, and let Prof. Hitchcock count them if he can with a low half-inch (say of 40°). I hazard the prediction that he will fail, and furthermore, that I will be enabled to count them with an inch of 45°. I am ready for such test in any shape that Prof. Hitchcock may suggest.

To get back to our "moutons." When I stated 30,000 as the limit for definition of the inch, I held a good large margin (to swear by) in reserve. This, as the 30,000 was "millions" as far as establishing my point was concerned, now that the readers of this journal have got used to the "30" I will add that I have no trouble in seeing the 40,000 band of a Rodger's test-plate, and most beautifully ruled they are. Here is an added strain of ten thousand on the inch, but it is as good as ever! (One of these days I must give your readers a description of an exquisitely ruled plate, up to 80,000 to the inch, and by the same talented artist.)

But to business. I have as fine a half-inch of 38° as I have ever seen, a glass selected with much care, having spent considerable time in making the selection, so as to have the very best; for, be it known, I *sometimes* (not often) have use for just this sort of glass. Well, now, this half-inch has shorter working distance, and, as a matter of course, less penetration than has the inch of 45°, while the last-named inch of 45° has much the *superior definition*.

Now, I am perfectly acquainted with the "points" of these two objectives, and can make either put its "best foot foremost," (hence the "handling" may be counted out.) Both are my own, and I can at will use the one that I prefer. Which would Prof. Hitchcock use under the same circumstances ? Or, leaving the Professor out, which would the reader select? Which glass would be "strained" the most?

Prof. Hitchcock reminds his readers that he has "already admitted to Prof. Smith that Tolles and Spencer are making wonderful glasses, etc." Yes, so he did, but he put a fence round them ; thus he adds "but for *resolutions only.*" When Prof. Hitchcock shall become as well acquainted with the work of Spencer and Tolles as are others, that fence will step down and out. I accept no loop-hole, but stand to the position heretofore assumed, to-wit, that the above named glasses are not equalled for "any and all work," as set forth in my last paper.

Now about Dr. Johnson's showing the 19th band as an opaque object in the year 1872 or thereabouts. Here is a legitimate chance to try on Prof. Hitchcock's conditions, to-wit: Were

Dr. Johnson's experiments with the 19th band published? (as were my own). Again, were they "subjected" to public examination?" (as were my own.) Let the Professor's rule work both ways. Meanwhile I now assert, and know whereof I affirm, that a Powell & Lealand 1-10th of 1872 cannot be forced to show the 19th band by lamplight. In general, I don't propose to accept anything done with the 19th band five years ago, as comparable with what is being done in 1877. Again, all who have seen the 19th band as displayed with the duplex and modified Beck illuminator, instantly admit that for wondrous delicacy of definition, the total absence of the "ghost" of a diffraction line, all other methods of illumination (by artificial light) are simply "nowhere." And further, that any attempt at photographing these wonderful lines has been but approximative. To be more definite, the best photograph of the 19th band yet made is simply a caricature. Nevertheless it is wonderful that Col. Woodward has succeeded as well as he has, and his photographs are the best that I have seen.

A word as to the mathematical tables. There can be no possible objection to these, *when* (it's my chance to say "when" now) they agree with *obvious facts*; they are objectionable when they are not in accord with facts. I stand ready to accept truth in any form ; ditto, to repel error. Now the facts are, that from a mathematical standpoint the limits of vision have been erroneously "set" again and again. I simply protest against fighting *facts* with mathematics—simply this, and nothing more.

In drawing towards the close of this article, I desire, as *apropos* to the question at issue, to state a little bit of personal experience, viz : About a year ago a gentleman, a professor occupying a responsible position, gave me a *carte blanche* to provide him with a medium power lens. He was particular to state that he was engaged in histological studies, and wanted the glass for such and similar purposes. In response, he was furnished with a 1-6th duplex. Last summer he visited me during his vacation and became my pupil. He was, of course, an apt scholar, and quick to learn. He spent with me weeks in acquiring the manipulations of this glass, and became quite expert. When he left me he said, " I have now learned *practically* what was done for me in the selection of the objective, which I now prize more than ever, and it is astonishing *how much* is due to handling. I had no idea of it." A few weeks ago he wrote me that he was doing finely with the duplex, and had succeeded with it in tracing structure (histological) that could not possibly have been seen with any low-angled glass. Further, that the very best makes of the low angles were constantly within his reach.

Now it is a fact that before I furnished this gentleman with the duplex, he was as fully committed in favor of low angles as Prof. Hitchcock can be; but, unlike Prof. H., he commenced the study of high apertures, and with the sure results above narrated. His address will be furnished to any one desiring it.

Prof. Hitchcock says something (I quote from memory) about "accepting defeat in a becoming manner." I desire most earnestly to advertise the fact that it is not, has not, been any part of my purpose to "defeat" Prof. Hitchcock, and as I have understood the gentleman from the very onset, he is not, nor has he been, in a position to suffer defeat. He simply wants the facts, and I have endeavored to furnish them, and if so be that any good has been brought about by this discussion, your readers are quite as much indebted to Prof. Hitchcock as to any one else.

In conclusion, I return Prof. Hitchcock my sincere thanks for the courteous and gentlemanly consideration he has ever extended me. It has been my intention to reciprocate, and with the hope that I have been in a measure successful, and with the kindest feelings towards Prof. Hitchcock, to you Mr. Editor, and to your many readers, I remain sincerely yours,

J. EDWARDS SMITH.

NOTE.—In your October number, page 132, an obvious error occurs. For "what was known as a high-angled 1-10th of 180° ten years ago," read "130° ten years ago," etc.   J. E. S.

DISCURSORY THOUGHTS RELATING TO THE USE AND ABUSE OF THE MICROSCOPE.

An Addrass delivered by Prof. J. Edwards Smith, before the Dunkirk Microscopical Society, Tuesday Evening, October 31, 1876.

To the Members of the Dunkirk Microscopical Society :

GENTLEMEN:—About twelve months ago a friend of mine and a brother microscopist, ordered from London a low-angled 1-4th inch objective ; my friend was engaged in histological investigations, and felt the need of a reliable medium-power glass. He had been told that wide-angled glasses, although very suitable for the purposes of the diatomist, were hardly suitable for the work he had in hand; he was told, too, that the thing needed was a glass of low or moderate angle, say from 50° to 70°, and one that had great working distance; in short, as the saying goes, he needed a good " working glass."

In due time the objective was received, and one very pleasant evening my friend called on me expressly to show the work of the lens. I remember that we had a very enjoyable time, amusing ourselves by looking over a variety of specimens suitable for the glass. In the course of the evening he exhibited a beautiful section of a human tooth, and placing the same under the 1-4th, called my attention to the "Nasmyth's membrane," described by Frey and others.

This was a matter of interest to me, for I had heard much of this "Nasmyth's membrane ;" I had examined human teeth time and time again with my own wide-angled glasses, but had never been able to discover the same. I was, in fact, somewhat curious about the matter. It seemed odd, to say the least, that my glasses, which would show me clearly and accurately, lines as close as 120,000 to the English inch, should fail to show me the existence of a membrane said to be about 1-6,000 of an inch in thickness ; and then again I had hunted with a low-angled 1-4th of my own without success. Hence it was with expectancy and interest that I put my eye to the tube ; I saw a decided thickening of the margin of the crown of the tooth, and was informed that this was the veritable "Nasmyth." But, said I, "is this thickened edge not the fault of the objective, and suppose that we examine the specimen with another glass?"

Removing the London lens, I substituted a wide-angled 1-6th of plus 180° of aperture, and on adjusting carefully to the thickness of covering glass, I soon had a splendid view of the preparation, by far finer than that shown by the London glass; and as I had supposed would be the case, I now had a nice clean edge in the place of the thickened one before mentioned. The superiority of the definition over that of the 1-4th was apparent; and the capacity of the 1-6th to bear the 1-4th inch solid eye-piece and amplifier—thus carrying the amplification up to nearly 4,000 diameters—was in a few moments demonstrated, but the Nasmyth's membrane was not to be seen under any amplification.

At this time it occurred to me that I might possibly *make* a membrane as described by the books, and as shown by the London lens. Turning the collar of the 1-6th so as to throw the glass considerably out of adjustment, and again focussing the object, I was much amused to find that I had the identical membrane, as shown by the imported glass, and also, that as the collar was moved, the membrane could be made thick or thin to suit!

The result of this little episode was simply this—my friend now owns and uses on *histological* work an American made 1-6th of 180° of aperture.

I have related, as briefly as possible, this little incident, and exactly as it occurred. I desire to say, however, that I do not deny the existence of "Nasmyth's membrane," described in the various works on physiology and histology. Nevertheless candor compels me to state that I have in times past made diligent search, and have failed to recognize a structure which is taught even in our elementary school books, as if it were a commonplace thing!

Months ago it occurred to me that the membrane referred to might only be found on teeth of the first dentition, and obviously would be found wanting on the tooth of an adult. This idea sent me to the dentist's office; such specimens were procured, but the "Nasmyth" still defeated my search therefor.

Some months ago I was engaged in teaching microscopy in one of our public schools. At that time I had the class in physiology in hand, and it was my especial purpose to show them *actually*, structures which were described and pictured in the text book used. I had not proceeded far before we ran against this " Nasmyth's membrane," and I was compelled to state substantially the experience which I have now related to you.

There is not a class of persons on earth less appreciated or more poorly paid than those engaged in histological investigations, and for their indefatigable perseverance I entertain a profound respect; but it has often occurred to me that as a class they have been content to work with very poor tools, and it is safe to say that a great deal of the ground already studied will have to be reinvestigated, and with better glasses—with a better knowledge of those manipulations requisite to the use of wide-angled lenses. Nor can the necessary manipulations be acquired in the ordinary routine of investigation ; it is an art of itself, and calls for special attention, requiring as much study as the art of manipulating a pianoforte or an organ.

It is a very commonplace remark, that, for the purposes of the histologist, low-angled glasses, of moderate definition, are the best suited. And yet, if we consult their latest works, we will find structures described that call for the best lenses of the widest angles. Take, for instance, the muscle sheaths, or the axis cylinders of the nerves described by Frey, in his last edition. Coming down to commonplace things, I affirm that the simple trachea of a bee cannot be studied with advantage with other than lenses of the widest angles.

There is perhaps no microscopic object more common than these tracheae. They are to be found mounted in the cabinet of every microscopist, and we have all of us read about them in the books ; we have examined them again and again, times without number, with objectives ranging from the inch to the 1-5th, or perhaps to the 1-10th. Now I ask you take a look at them on my account, and if possible inform me how these tracheal coils terminate ; what is the structure at their *very last end?* What is the least diameter of these terminal coils ? Here you have a problem in hand, to which, for exquisite intensity of definition, or nicety of manipulation, the resolution of the Moller or Nobert plates are child's-play matters ! And if you should be desirous of testing your lenses as to their capacity for histological work, the trachea will be found to be one of the very best test objects in existence, and will surely furnish you valuable ideas as to the relative value of wide and low angled glasses for the purposes of histological research.

Not long since one of our most skilful physicians sent me a sample of urine from a child supposed to be suffering from a disease of the kidneys. It was a perplexing case, and the medical gentleman in charge hoped that the microscope would give some light on the case ; when examined with a low power 1-5th of 70° aperture, the specimen appeared in every respect healthy ; but on further examination with a wide-angled glass, and with an amplification of nearly 4,000 diameters, it was found to be literally swarming with vibriones. There were billions upon billions of them in every field examined, all of which were totally invisible to the low-angled 1-5th.

We often hear the remark, that wide-angled glasses are just the thing for the display of lined objects, surface markings, diatoms, etc., but that owing to their short focal length, and limited working distance—the trouble attending the adjustment of collar—and in general, the difficulties pertaining to their use, that they are unsuited to the purposes of the histologist ; while on the contrary, low-angled glasses of greater working distance requiring no skill in management, are the tools with which the real work of the microscope has been, and will continue to be done, and such are fondly termed good, honest and reliable " working glasses."

I can never listen to this line of argument without entertaining the suspicion that sloth and inactivity lay at the bottom. We never hear astronomers complain of the care they are compelled to use in instrumentation ; on the contrary, they pride themselves on the accomplishment of being able to work instruments requiring a great amount of skill and precision in manipulation.

The objection of short working distance originated years ago, when German pathologists were in the habit of using common window glass to cover their mounts, and at that time the extremely thin glass, now so easily obtainable, was unknown. All this is now changed ; the widest angled lenses known giving all desirable amplification, will admit the use of covers 1-50th of an inch thick, while covering glass measuring .003 can easily be procured.

A vast amount of work *has* been done with these "honest and reliable working glasses," and, as I have before said, will have to be done over again, and this revisory work is now in progress. But while there is some excuse for the investigators of the past, who used the best instruments then obtainable, what shall we say of those of to-day who persistently refuse to avail themselves of the wonderful progress the optician is able to demonstrate in this centennial year ?

I have advanced the idea that working pathologists were too often content to work with poor tools. I made that remark in good faith, and believe tnat it will be found to be trne. Nevertheless the pathologists of the past, or even of the present day, are by no means wholly at fault, when we remember that the entire corps of observers to whom we are indebted for all that has been taught through the aid of the microscope, were men who were regularly engaged in the respective duties of their several professions, and that most of them have filled professional chairs in various institutions of learning. These men, in studying the use of the microscope, each and every one of them were compelled to dig and trench for themselves ; the instrument which to them was as all-important as the compass to the navigator, was far, very far, from being as true and reliable as "the needle to the pole." Once launched on the unknown, but to them, fascinating sea of scientific investigation, every spare moment that could be snatched from their regular engagements was devoted to its exploration. It was a tiny sea, within a microcosm, replete with interest, and capable, as if by the touch of the optician's wand, of boundless expansion !

These self-abnegating, self-sacrificing men fully accomplished their task. It was a labor of love, and the results they obtained were given to the world without money and without price.

Among the investigators of our own day, I will refer you to one whose untiring industry has earned for him a world-wide reputation. I allude to Dr. Lionel S. Beale, F. R. S. Dr. Beale, as you are aware, has been contemporary with, and the supporter of, Dr. Carpenter's heretofore universally received opinions as to the relative capacity of wide and low angled objectives. Dr. Beale, in his work, "How to Work with the Microscope,"— I quote from the third edition, page 7 —says as follows: "For ordinary work it will be found inconvenient if the object-glass when in focus comes too close to the object. This is a defect in glasses having a high angle of aperture. * * Glasses with a high angle of aperture admit much light, and define many structures of an exceedingly delicate nature, which look confused when examined by ordinary powers. For general work I recommend glasses with an angle of not more than 50° to 100°

"Mr. Ross has lately made glasses having an angle of 170°, which are valuable for investigations upon very delicate and thin structures, such as the diatomaceæ; but such powers are not well adapted for ordinary work. The importance of arranging the object very carefully, and the necessity of paying great attention to the illumination, render these glasses inconvenient for general observation. The penetrating power of glasses with a low angle is much greater than in those of a high angle of aperture, so that exact focussing is much more important in the latter than in the former."

This, in its popular acceptation, is the square doctrine first advanced by Dr. Carpenter, and it has suffered no loss by its filtration through the mind of Dr. Beale. It is the doctrine which has been generally received and accepted. It is the dogma which ten years ago I thought ought not to be true, and of which to-night I stand before you the prince of sceptics !

In the quotation which I have presented, there are two dominant ideas; the one displays the *inconvenience* attending the use of wide-angled glasses, and to this objection I have already paid my respects in due form; the other presents the theory of " penetration " as originally advanced by Dr. Carpenter, supported by Dr. Beale, and generally endorsed by microscopists of Europe and America.

I have not the time this evening, nor would I weary your patience by the discussion of this dogma of penetration, any further than to admit that it was true ten years ago, when first announced by Dr. Carpenter, but can have no possible *present* force in reference to the wide aperture glasses of to-day. I am also quite prepared to grant that a spectacle lens of sixty inches focus needs less skill in management than a duplex 1-10th of plus 180° of aperture!

After all, I get considerable comfort out of the quotation already presented. Dr. Beale says, "that glasses with a high angle of aperture admit much light, and define many structures of an exceedingly delicate nature, which look confused when examined by ordinary powers." Notice the wording; he does not say diatoms. He does say, "many structures of an exceedingly delicate nature, which look confused when examined by ordinary powers." It is a candid and manly admission, and I honor his candor. It is the truth, the whole truth, and nothing but the truth; while for brevity of form, or clearness of expression, these half dozen words leave nothing to be desired beyond what is here so forcibly stated; and I beg of you not to lose sight of the fact that if this was true of wide-angled glasses made in 1865, it is *equally* so of those of 1876.

Another fact will bear to be borne in mind. It is this : when Dr. Beale wrote in 1865, he was very far from being expert in the use of wide apertures. I will read to you further from the same paragraph : "In order to adjust the object-glass, it is first arranged for an uncovered object; then any object covered with thin glass is brought into focus by moving the body of the microscope; next, the ring which carries the third lens is screwed round until any particles of dust upon the upper surface of the glass are brought into focus. The glass is then corrected for examining the covered object which may be brought into focus." It will be seen that the method of adjustment adopted by Dr. Beale is precisely the same as that used by many for measuring the thickness of covering glass, and when the milled head of the fine adjustment is properly graduated, and the "run" of the fine adjustment is known, the plan is a convenient one, and will give measures tolerably correct; but for the purpose of adjusting an object-glass it is obviously faulty, and would in almost every instance defeat such wide angled lenses as I am in the habit of using. Dr. Beale's instructions as to the adjustment of wide-angled lenses must be taken "cum grano salis," and will indeed be more honored in the breach than in the observance.

It is with pleasure that I now turn to page 217, and read as follows: "Besides extreme minuteness in mere size, extreme tenuity or transparancy may interfere with the definition of an object. Now the greatest difference is observed in object-glasses in this particular. The *best* object-glasses will define *clearly* and *accurately*—mark the words," clearly and accurately" —bodies which, from their transparency, are quite invisible under objectives only slightly inferior to the first."

Now the question arises at once, Where shall we find these best object-glasses. which define so clearly and so accurately such delicate and transparent objects? Referring again to page 7, Dr. Beale responds thus: "Glasses with a high angle of aperture admit much light, and define many objects of an exceedingly delicate nature, which look confused when examined by ordinary powers."

The Doctor having thus squarely settled the question, most kindly goes on to tell us where we can get these wide-angled glasses, and I read right along, as before quoted: "Mr. Ross has lately made glasses having an angle of 170°, which are valuable for investigations upon very many delicate and thin structures," etc., etc.

It is especially a part of my purpose in appearing before you this evening to show you "clearly and accurately," and by the aid of glasses of American manufacture, which for width of aperture have never been excelled, some of these extremely thin and transparent objects, and thus prove to you the truth of Dr. Beale's assertion; and relying on the correctness of the old adage that "seeing is believing," I can hardly fail in the demonstration of the accuracy of Dr. Beale's position.

Pardon a momentary digression, and allow me to turn to page

216. Here I read: "If any one makes out new points of structure by any new method, all that such an authority who differs has to do is to state that he has not been able to see the structure described so and so. *Authority* too often denies the existence of what it has itself been unable to see. Many authorities deny the existence of what they have not seen, while they have not taken the pains to try the only method of demonstration by which the appearances in question could be seen."

Now we perceive that the Doctor is in a complaining mood. The root of the matter was simply this: Dr. Beale paid great attention to instrumentation; the first 1-25th and the first 1-50th objectives in existence were made for him; he was not very particular as to their angle, although it is noticeable that he claimed for them "plenty of light." But the fact is indisputable that his instrumentation has been in advance of that of the London microscopists. Dr. Beale was in fact in a position very similar to those of the present day, who, conscious of the superiority of their glasses, are disposed to regard the blind " authorities " referred to as unworthy of special consideration.

I desire in this connection to call your attention to the history of the diatom of which we have all of us heard so much about during the past nine years. I refer to *Amphipleura pellucida*. Who of us here cannot recall the time when the existence of striæ on this shell was stoutly denied? We can remember, too, when Dr. Woodward settled this much vexed question by producing his photographs displaying the striæ on what he was pleased to call " this well-marked diatom." And now that these markings are no longer to be questioned, as exhibited on dry mounts, and by the employment of monochromatic sunlight, full eighty per cent. of our American observers fail to exhibit " this well-marked diatom " by lamp illumination, even when dry mounted. I have a balsam mount of this shell with me, and hope to show you the lines of No. 20 of the Moller plate this evening before we part, and also the 19th band of the Nobert test plate, and by the aid of American made lenses of medium power.

Let me call your attention, for a few moments only, to another subject. We often hear the diatomist spoken of in terms,

almost, of contempt. They are too often regarded by the histologist as a class of observers who use the microscope as a mere
plaything; and the fact that the diatomists are not altogether
agreed as to the structure of some of their favorite shells, is
often used as an argument to show the folly of studying the
diatomaceæ at all !

All this, my friends, is sheer sophistry. The study of the diatomaceæ is as legitimate as that of any other branch of the
science of biology, and the labors of the diatomists have not
been for naught; it is to them, and to their constant demands
on the optician, that we are indebted for the wonderful
improvements which have been made in object-glasses; and I am
bold enough to tell you that skillful diatomists can tell you as
much concerning the structure of a diatom, as can the patholo-
gist—equally skilled—inform you as to the structure of a blood
corpuscle !

But to the student, to those who desire to *prepare* themselves
for advanced work, the study of the diatomaceæ cannot be neglected. No line of practice has yet been discovered that will
teach the student the use and management of his tools, that
can at all compete with the superior claims of these minute organisms. It is said that "adversity tries us and shows up our
best qualities." These little shells, too, will try the would-be-
manipulator, and, like the country judge, show up his worst
qualities.

It was not my purpose to enter at all into the details concerning the use of objectives; but it will perhaps be well not to let
the opportunity pass without alluding to the fact that a wide-
angled glass requires totally different management, in some
respects, from those of narrow apertures.

Those who have been accustomed to the use of the low angles
will, on a slight acquaintance with glasses possessing wide apertures, almost invariably assert that the latter do not give as
good results, when worked by central, of centrally disposed
light, as they are accustomed to get from the former. A
moment or two devoted to the consideration of the situation,
will, I think, furnish the key thereto.

When we use a narrow-angled objective, the oblique or lateral rays are per force excluded. The objective will not receive them, and there can be no possible doubt but that such work is being done with centrally disposed light, such as the observer desires; there is no special manipulation or management necessary to secure this, for the objective will admit centrally disposed rays, and none other.

But in the handling of a wide-angled lens the conditions are essentially changed, and this change of condition involves in turn a change of handling, and of management.

Hence it occurs, that in using a wide-angled glass by central or centrally disposed light, some arrangement must be provided for the purpose of shutting out the lateral rays; these the glass will admit, and in default of the provision referred to, will steal in, cause interference, and defeat both observer and objective. To shut out these lateral pencils would seem eminently the business of the usual diaphragm box.

This provoking little piece of apparatus, this diaphragm box, is supplied with almost every stand in use, and is as faulty in operation as human ingenuity in construction could devise; in general these boxes are furnished with a shutter pierced with openings of various sizes, and placed from one-half to one inch below the object-carrier of the stage. Now, suppose we are using a wide-angled lens, and being desirous of central light only, we attempt the use of the smallest opening of the shutter. What now is the result?

It is as follows: the pencils of light enter the small aperture, emerge and diverge within, and fill the box with simply light of low intensity. It has now become a washed-out, wire-drawn light, lacking force, and, I may say, velocity; the objective now in turn receives this much-abused illumination, not only the central pencils thereof, but lateral ones also; and the working angle used will be determined by the depth and diameter of the diaphragm box.

A proper diaphragm for general use—not diaphragm box, for the less of the *box* the better—is a plate pierced with a central opening of about the size of a large needle, and so mounted as to approach the lower edge of the mount closely as possible without suffering actual contact. To construct this little piece

25 Microscopy.

of apparatus will not make any very severe demands on your mechanical skill.

Those of my hearers who read the London *Monthly Microscopical Journal* will remember that the capacity to display the markings of *Navicula angulata* by central light, was considered a feat only to be performed by low-angled glasses of superior excellence. Some of you, too, may have read the report of a certain biological committee who met in Philadelphia last year for the purpose of testing the object-glasses of several makers ; one of the tests used was the exhibition of the markings of this same dry mounted *Navicula angulata* by central light.

I was simply astounded when I read the report, in public print, of this biological committee, and learned that in their hands that *Angulata* had defeated a wide-angled American 1-10th, and I immediately repeated the experiment, using a similar 1-10th, by the same maker, and worked with a diaphragm plate perforated with an opening, say, 1-200th of an inch in diameter, and placed almost in contact with the under surface of the mount. The result was amusing enough ! I instantly had the markings " clearly and accurately " defined ; the problem was, indeed, to avoid seeing them ! In fact nothing short of sheer intention, or, what is worse, bad manipulation, could have defeated the objective. I then repeated the experiment, using this time a balsam mounted *Angulata*—the No. 11 of the Moller plate—and was almost instantly rewarded, as in the previous instance. I am prepared to again repeat this experiment this evening should you so desire.

Again, the converse of what has been said is also true, namely in working with a wide-angled objective, and with oblique light, it is important to shut out the central pencils. About a year ago, I devised a simple little instrument, which gave access only to a narrow wedge of oblique light, and thus gave added force to the definition of the objective—and this " oblique diaphragm " was permanently fitted to the stage of my stand.

The most efficient instrument of this nature, however, is " Wenham's reflex illuminator," an ingenious accessory, and so contrived as to shut out all rays less than 41° interior, which, of course, has emergence at 90° into air. The reflex illuminator, which was designed by its inventor to give a dark field, becomes, when used on balsam mounts, with American object-glasses, having balsam angles ranging in the nineties, not a *reflex*, but a *direct* illuminator, and a most efficient aid to the definition of the lens. By it I am enabled to show the transverse striæ of No. 20 of the Moller plate so clearly that any old lady who can read her family bible could hardly fail of recognizing the striæ on "this well-marked diatom." I have the instrument with me, and shall be happy to show you its work on the Moller plate. *

It was mainly the purpose of these desultory remarks to call your attention to the importance of instrumentation ; absolute perfection has not been, and never will be obtained. There never has been, nor will there ever be made, an instrument of precision that does not embody inherently some radical weakness, some dangerous fault. The practical astronomer is not only early taught the nature and use of the various instruments to be found in the observatory, and made acquainted with their several imperfections, both optical and mechanical, but he is compelled to acquaint himself thoroughly with the methods employed to eliminate these imperfections. Having thus perforce of his preliminary study acquired a thorough perception of the Scylla and Charybdis which environ the use of instruments, it becomes in turn a life's study to further remove existing difficulties, or to provide better and more competent means of compensation therefor.

I have no doubt but that it would be interesting and instructive to detail some of the exquisitely precise methods employed by the astronomer to detect and compensate for the unavoidable errors of instrumentation, but time forbids. I will, however,

* Immediately after the introduction of the "reflex," modifications, by changing the angle cf the facet, were made by the London opticians, so as to adapt the instrument to their variously-angled objectives, similar modifications were also made here in this country, by Messrs. Tolles & Spencer. Three years ago the writer consulted a skilled artist with the view of making the "reflex" in three separate mountings, changing the angle of the facet in each. Nevertheless. the idea is Wenham's. The mere matter of changing the angle of the facet would naturally occur to any one using the instrument—J. E. S. January, 1878.

notice the nature of one or two difficulties which may be said to infest the observatory.

We are all familiar with the purpose for which the transit instrument is employed, namely, to observe the transit of celestial bodies across the meridian in which the transit is adjusted. To accomplish this, it is evident that the instrument must be permanently placed in that meridian; but, unfortunately, the block of stone has yet to be found solid enough for the purpose. Hence it is usual to ascertain the direction and amount of error and thus correct the results obtained by observing with the instrument while out of the plane of the meridian. The altitude and azimuth instrument, too, requires frequent and careful attention. To be assured that its object-glass may swing in a vertical plane, observations are nightly made by observing a distant star directly, and immediately afterwards its image as seen in the mercurial horizon. By this means a truly vertical line millions of miles in length is obtained.

Reflecting instruments constructed on the model of the sextant, are cross-questioned by observations on east and west stars, and errors due to eccentricity thereby avoided.

Not only do these painstaking men study their instruments, but they in like manner study themselves, and their capacity to observe is cross-questioned in the severest manner. It is to them well known that the observations of some first-class observers are constantly affected with a plus sign; while, again, those of other equally good observers are affected in a contrary direction. This individual condition is called " personal equation," and has to be thoroughly known and compensated for; and this you must bear in mind, that the amount of error due to personal equation is always comparatively small, and as a consequence is only recognized in the observations of professional experts, trained to machine-like, impassive regularity.

My initial remarks, in which the " Nasmyth's " membrane of the books was discussed in a somewhat discursory manner, were intended as an example, and to give you an illustration in reference to a radical evil, which has in times past tainted the observations of many who use the microscope—I refer to diffraction lines and diffraction borders. To avoid these spectral, illusory appearances, is a consideration of the very first importance— requiring, too, the employment of the finest object-glasses, and that technical skill in the adjustment thereof, which is only acquired by long practice; and to this end the study of the diatomaceæ will be found in the highest degree advantageous. Moreover, one must devote a good deal of study to this ghostly parasite, which too often infests, and to a greater or less extent defeats the unsuspecting observer. And here again the little diatom will render yeoman's service.

Some months ago, Dr. Woodward, to whom we are all so much indebted published in the London *Monthly Microscopical Journal*, for the express purpose of displaying these illusory appearances, a lithograph from his photograph of *Frustulia Saxonica*. The objective used by Dr. Woodward was an American 1-18th, and was the same glass with which his inimitable photographs of the 19th Nobert band were accomplished.

I took it for granted that the lithograph of *Frustulia Saxonica* was made with the glass intentionally placed out of adjustment, and for the purpose of creating these particular lines. A copy of the London journal is placed at our service this evening, and I invite your inspection of the plate referred to. The whole shell of the diatom will be seen badly distorted, the striæ badly defined, and the diffraction lines are immense!" *

---

* I may here remark that only a few hours ago my attention was called to a report of a meeting of the London Microscopical Society, held April 5, 1876, and reported in the May number of the London *Microscopical Journal*.

Alluding to the photographs of Frustulia Saxonica, Mr. John Mayall said: " Every one who is familiar with the Frustulia Saxonica, photographs of which Dr. Woodward sent in illustration of his paper in December, knows it to be one of the most difficult test-objects — a diatom that ranks next to Amphipleura pellucida. That particular form of Frustulia is one that I have rare y seen resolved, except by lenses of the highest excellence. I consider Dr. Woodward's photographs of it as in every way most remarkable, evincing first-rate skill brought to bear on one of the finest known lenses." It is hard to tell which is the most "remarkable," the photographs, or this statement of Mr. Mayall's.

As there seems to be much difference of opinion as to these representations of Frustulia Saxonica, both at h me and abroad, a word or two may be admissable. When the lithographs first made their appearance in the London *Monthly Microscopical Journal*, one of our most talented experts wrote

Now, after seeing and learning what can be accomplished with a superior objective, when out of adjustment, it will be instructive, and I shall beg your permission, by the employment of a superb glass, accurately adjusted, to show you a similar, but more difficult *Saxonica*, as illuminated by the most oblique beams my extremely thin stage will admit.   Thus handled the resolution of *Frustulia Saxonica* becomes one of the most charming and fascinating objects that can well be imagined.   We shall thus see the frustule without sensible distortion the striæ displayed with such exquisite beauty of definition as must command your admiration, and minus, too, the suspicion of a diffraction line !

The attention of Andrew Ross, while examining Podura scales with glasses of his own manufacture, was called to the thickened edge, the " Nasmyth membrane " so plainly seen with the non-adjustable glass he then employed, and to him are we indebted for the collar adjustment now so common to all first-class objectives.

Now, my friends, *good object-glasses*, like astronomers, have their " personal equation."   They alone are the ones to be most affected by collar adjustment ; an ordinary glass, furnished with compensating screw, is scarcely superior to an ordinary non-adjustable lens ; one may turn the collar through its entire range, without sensible or corresponding change in definition.

me saying that he was surprised that Dr. Woodward should have allowed the publication of prints giving such indifferent ideas of the work of American lenses.   I replied, that as I then understood the mat er, it was Col. Woodward's intention to show the work of the objective purposely placed out of adjustment, and I so continued to think at the date of the above lecture, a fact obvious to the reader.   I also suggested that the lithographs might not fairly represent the photographs ; that if it was the case, Col. Woodward would surely make the fact known.   His silence, however, authorized the inference that the original photos had not suffered at the hands of the lithographer.   I th nk I can now safely affirm that the general opinion is that one of these lithographs was intended by Col. Woodward to give a correct idea as to the appearance of F. Saxonica when properly resolved, *i. e.*, that it might be contrasted with the others then presented. Be all this as it may, I am prepared to assert that *no one* of said lithographs gives any idea of the proper resolution of Frustulia Saxonica.— J. E. S., January, 1878.

On the contrary, objectives of the widest apertures, and capable of yielding intense definition, require the strictest attention to their adjustment. *

Nor is this all. It is quite possible to accurately adjust a first-class objective, and, nevertheless, in this very act defeat the maximum performance of a first-class glass !

When working with oblique light, the maximum performance of a superior glass will be obtained at maximum aperture ; if this maximum aperture obtains, with the systems "closed," as is often the case, it is then manifestly our business to use covers of such thickness as will "correct" the objective *at* "closed," and to work such a glass over thinner covers, requiring the lens to "correct" at, or near "open point," would surely defeat its best performance.

It being possible that we may have time in the course of the evening to practically discuss this point, I have brought here an objective which, in any position of its collar, has plus 180° of aperture ; its balsam angles are at "closed," 97½° ; at half way between "open" and "closed" 95° ; at open point, 85°. It will be an easy matter to demonstrate that this glass is defeated by the thin covers of the Moller plates, over which the glass will "correct" near to "open point," and conversely, of the decided increase of definition obtained (and angle also) when worked through a supplemental cover of sufficient thickness to cause adjustment with the systems at, or near "closed."

I have brought here another, and a totally different objective; this glass has a constant balsam angle of 100° through nearly the whole run of its adjusting screw. Manifestly, when using this glass, there need be less attention paid to the thickness of cover, a fact which is demonstrated in practice.

---

* In an article contributed by Mr. F. H. Wenham—see London *Microscopical Journal* for March, 1876—I read, as follows : "The adjustment seems to be a stumbling block for those advocating an extra immersion theory. We have now in use thousands of serviceable immersion object-glasses, capable of defining most tests, and which have no adjustment, as they are set for an average thickness of cover. They answer well, because in the *immersion* system the errors of cover aberrations are nearly eliminated, and with a balsam intermedium they would be inappreciable." (1 1 1)

-- J. E. S.

These two objectives were constructed expressly for me, and were especially intended for conjunct use—they are in truth, companion glasses; either of them will display the 19th band by oblique use of artificial light, and also the markings of *Angulata* by central lamp illumination. Of the two, the first-named has the greatest working distance, and, for reasons already given, will work at *maximum* performance over covers 1-50th of an inch in thickness, and in the hands of one thoroughly conversant with its functions, it is, perhaps, the most generally useful glass.

I have here still another objective of plus 180° of aperture. This glass was intended to be a compromise, and to serve to a great degree the purposes both of the 1-6th and the 1-10th before mentioned. The glass I now refer to has at " closed " a balsam angle of 97½; at open point, 92. The compromise is thus apparent: this 1-6th was made to my order for the gentleman who fills the executive chair of this society. It asks no favors of M. Nobert, nor of a centrally posed *Angulata*, and when one is confined to a 1-6th alone, a glass of this construction will prove of great value.

These conditions, which have still greater force when we include the work of various opticians, need attention and study in default of which you may expect diffraction lines and diffraction borders; and, perchance, our old friend, the " Nasmyth," may put in an appearance!

I repeat, that it was the dominant purpose of these desultory remarks to call your attention to the importance of microscope instrumentation, and to the manipulations connected therewith. To the fact that it is a study, a profession in itself, and an accomplishment that must be fully mastered at the outset, before one can investigate with precision, or have claim to the confidence of others. In point of fact, the most expert manipulator will constantly have something to learn—there is no landing place where one can stop and rest. The world progresses, and so should the microscopist—*pari passu.*

I take it for granted that nothing I have said will lead you to suppose that I am opposed to the legitimate use of low-angled objectives; nothing could be further from my purpose. When properly used, for preliminary examinations, they are *convenient* and useful, but for advanced work they should be abandoned, and in favor of more *inconvenient* glasses of wider apertures.

A wide-angled lens, incapable of receiving and utilizing central beams, is a faulty and undesirable objective ; it is, in fact, but half a glass. And on the other hand, a low-angled objective, incapable of receiving lateral beams, is, in my opinion, equally undesirable, unless, indeed, when we are prepared to sacrifice force of definition to convenience of handling.

It must be admitted that there are functional and characteristic differences in the performance of the two classes of objectives ; for instance, a wide-angled glass of relatively shorter focal length will not see so far around an object as will one of narrow aperture and longer focal distance. This, as well as other considerations which might be mentioned, offer no rebutting testimony to the statement already advanced, that the principal advantage presented by the low angles is their convenience in manipulation.

In conclusion, I have to thank you for your kind attention to the rambling remarks I have had the honor of presenting. During the few short hours that I have been in your city, the earth has not had time to make one rotation ; but I have had a plenty of time to experience and enjoy a generous hospitality. I am now ready to show you the promised objects, and shall be happy to see what *you* have to offer. You will find me ready, willing and ardent to be taught, and in microscope matters be assured I am as absorbent as a sponge.

A CHAPTER ON ELEMENTARY PHYSICS.

(Dedicated to Prof. Hitchcock.)

Part First — The Sun.— " The luminous orb, the light of which constitutes day ; the central body around which the earth and planets revolve ; a celestial body which can be seen almost any fair day through a piece of smoked glass."

I beg my friend to read the above little monograph on elementary astronomy carefully. Assuming his compliance with this request, I go on to say that just after *sundown* this evening, my attention was called to his really kind notice of myself, as well as my Dunkirk lecture. Allow me to assure you, my dear professor, that although we have been pulling at opposite ends of the rope for some months, the generosity I have experienced at your hands has been vastly in advance of what I could have returned, and has placed me in position to receive any criticism that you might offer with the best possible grace.

I now propose to jerk *my end* of the rope in a manner that will make things lively with you! The fact is, when I get after you with a " sharp stick " you become unusually brilliant with the jerk responsive! I like it.

I admit that you are one " who *demands* accuracy of expression in every branch of science "—" that the scientist has *no right* to say what he does not mean, and he has *no right* to presume that his readers will understand him if he uses incorrect language." I therefore have placed the above little monograph on elementary astronomy at your service, and in timely season, fearing otherwise you might be led astray by my use of the word " *sundown* "—you will see the point. It's true that Joshua of old commanded the sun and moon to stand still, and the story is found in a book generally admitted to be a classic, but —no matter.

Now, professor, I fear that you do not practice what you preach; for I read (page 92), " if this question were put to a thorough physicist   *   *   *   knowing nothing about balsam or fluid mounts, etc." Isn't there an " impossibility " here? A thorough physicist knowing nothing, etc. There's a screw loose somewhere.

Right on your next line is another jumble. *Here is* a want of accuracy or what not—which is it? You say " Must the scientist.be led by the nose," etc. If this expression *is* accurate, I want you to tell me just how this kind of a thing is accomplished; give me the p-r-a-c-t-i-c-a-l details, just as it would be done in the flesh, including the " adopt " and the " teach " —before an audience, too, without " apology "—and the why of it. " It's *the right and duty* of every man of science to ask this little question *whenever it occurs to him.*" " The scientist has *no right* to say what he does not mean." Let us have the facts.

Part Second.—180°, plus 180°, 180° + 45.° It is true, my dear

sir, that in my Dunkirk lecture I did use the words "plus 180°
of aperture." Now that select and intelligent audience (com-
prising physicists who *did* know something about "balsam
angles,") so far from demanding an "apology," generously and
unanimously gave me a vote of thanks. Curious, wasn't it? I
remember well my delight in there meeting the veteran Dr. J.
W. Armstrong, Principal of the Fredonia Normal School, and
one of the leading educators of your state; another physicist,
too, who *can* talk intelligently as to "balsam angles," and who
has made the apertures of objectives an especial study, and
who afterwards became one of my most valued correspondents.
Nor did we have any quarrel about the "plus 180°." Most of
the audience had read about Joshua, and doubtless would not
baulk at such expressions as "sunrise" or "sunset."

After all, professor, admitting that the "plus 180°" might
have courted some such gentlemanly criticism as it finally got
from you, I reckon I was as near right as Joshua. But before I
can lift you over this stile, I must scoop you up! get you
together in some shape, so you can be handled. First of all,
you seem to put the 180°, plus 180°, 180° × 45°, all in one boat
together; you regard them as synonyms. And, secondly, I
have to learn some things from you.

Messrs. Tolles and Spencer you denounce because they mark
their objectives 180°, which you say is not only "impossible,"
but "absurd," and that one of these gentlemen (you don't say
which) attempts to lead the scientist "by the nose." And you
further say that one of them (I don't know which, again,)
"might as well add 45° to said 180°."

Well, here is material enough to commence on. Now, my
dear professor, you maintain it to be the "right and duty" even
of every man of science to ask this little question whenever it
occurs to him the "why" part you know). That I have been
engaged in a controversy with you for months, demonstrates
my claim as a "man of science." You can't dodge that, and
the "occur" part is present and up to the high-pressure notch.

Now, I want to know "why" it is that 180° of aperture is
impossible; "why" that plus 180° of aperture is impossible and
absurd. We have your assertions, *but minus the ghost of a demonstration.*

Why do you accuse Messrs. Tolles and Spencer of perpetrating acts that are absurd and impossible; and *why* do you venture to hint that one of these gentlemen attempts to " lead scientists by the nose," (whatever that may mean). The names of Spencer and Tolles are revered by American microscopists, and their unrivaled efforts in the improvement of object-glasses have won for them a world-wide reputation. The chances are that they know more of microscope optics than you and I put together; and these are the men who mark their objectives " 180°." *Why* should there be " a law to prevent it ?"

Now, professor, if this 180° is impossible and absurd, will you kindly inform the readers of this journal *what* figures Tolles and Spencer ought to engrave on their wide-angled objectives in place of the awful " 180°." Will you be kind enough to name the extreme angle immediately adjacent, but not *contacting* the impossible and the absurd ?

Again I read, " plus 180° does not mean 97° balsam." Did I assert this ? Now, I ask *what* does 97° balsam mean ? Let's have it, and the "why" thereof. Unless you can tell me *exactly* what the 97° balsam angle *is*, I shall not take any stock in your above assertion.

Once more, you say that " true apertures can be measured and definitely stated." I wish that you would tell Mr. Wenham how the thing is done; he tried for a whole year to measure one of Tolle's objectives, without getting any two results alike! But I am after *you*, not Wenham. I desire to learn from you precisely what " true aperture is." When thus taught, then I desire to know by what physical process the same can be measured—" definitely" you know. "Accuracy" is the thing we scientists want.

Now, my dear sir, I call on you to answer all of these interrogatories, and when you shall have proven that 180°, plus 180°, or 180° × 45°, are one and the same thing; when you shall show that Messrs. Tolles and Spencer are asserting absurd and impossible things, and more, are trying to lead scientists " by the nose;" when you shall prove that a thorough physicist can know nothing of balsam or fluid mounts; when you shall have demonstrated that I should have apologized to my Dunkirk audience ; when you shall have proven what 97° of balsam angle is ; ditto as to "true aperture," and how to "definitely measure" the same--then and not until then will I take the " plus 180° " under advisement. Meanwhile we will be good friends, and with Gen. Grant say, " Let us have peace." Don't omit reading the reprint of Wenham's article on angular aperture, pages 74 and 75.

CHOICE OF OBJECTIVES.

Ed. Am. Jour. Microscopy.—It was with real pleasure that I read the article entitled " Dr. Carpenter on Angular Aperture," by W. G. Lapham, Esq., in your May issue.

First of all, it is very satisfactory, from the fact that Mr. Lapham speaks *from a practical standpoint.* The gentleman is an entire stranger to me, but the fact that his conclusions are drawn from his personal, and, I doubt not, protracted experience, is evident from the ability with which he handles his subject.

Your readers are well aware that I have often complained that I have been compelled to fight theory with practice and that no effort on my part sufficed to induce my opponents to abandon their theoretical ideas, and to examine practically as to the truth of the positions advanced by myself. In Mr. Lapham, however, I recognize a co-worker, and although the results he arrives at are not in perfect coincidence with my own, I find nothing to criticize, but am simply desirous of further comparing notes with the gentleman, in the hope that something to the advantage of microscopy may be developed.

I propose to present a few thoughts for Mr. Lapham's consideration, requesting the gentleman to give them whatever study and attention he may elect, and to advise your readers as to the conclusions he arrives at; and, as before intimated, nothing that I have to say ought to be taken in the light of a criticism of his really excellent paper.

First, I desire to ask if Mr. Lapham does not attach too much importance to the so-called " penetration." Is there in *esse*, any such thing as "penetrations?" Perhaps it will be well, first of all, to settle this point.

Now I hold as follows : We will take, for example, two glasses, both possessing the same amplifications and the same apertures ; now of these two glasses, the one having the greater working distance will exhibit structures situated slightly out of the precise focal plane, or what amounts to the same thing, will be less susceptible to slight changes in the focal distance, or again, to use the words of Dr. Carpenter, will have the greater penetration. In extreme cases, the item of aperture may be disregarded. For instance, as stated in my Dunkirk lecture, a spectacle lens of sixty inches focal length will be endowed with greater penetration than a Tolles duplex objective.

But in the Dunkirk address, I did not avail myself fully of the demonstration afforded by the spectacle lens, but now invite Mr. Lapham's attention, thus : If it so be that with the spectacle lens I am enabled to see with perfect clearness of vision objects across the road, it will surely occur that when I attempt to read fine print, that the lens will defeat me. I will, to be sure, see the lines of the print, will recognize the contour of the book, and larger objects in immediate proximity, etc. Here is an example from everyday life, to-wit : The glasses suitable for reading will not answer for observations at a distance. True it is that with such glasses, with which we read with ease, we may also see the general forms of things across the road, yet they suffice not to render clearly the *details* thereof. Hence we have all of us observed that many persons wearing convex glasses often use two pairs of spectacles, one for near and one for distant vision.

Now in this everyday case we are taught a lesson, that, from some cause or other, has been a slow one to acquire. First "penetration," in its naked aspect, is simply a function depending upon working distance ; and secondly that "penetration," *unless accompanied with a certain amount of definition, is practically worthless.* I ask Mr. Lapham's close attention to this point, one that has been terribly overlooked.

Now, if Mr. Lapham's one-sixth of 180° happens to just "turn the corner," *i. e.* have balsam angle, say, about 85°, I would feel sure that a little observation would lead him to the same opinions I have arrived at, and I only fear that he has chosen one of the highest balsam angles, in which case he will have to fall back on his four-tenths. Nevertheless, let him, if he will, try the experiment, and report as to whatever is gained in point of penetration by the use of his one-forth of 50° over and above what can be obtained by the use of the one-sixth, or the four-tenths. The point I have to make is this—the one-fourth of 50° will have the greater penetration (so-called, like the spectacle lens) *but will lack definition*, to the end that more is lost in the latter element than is gained in the former.

I state as a matter of fact, but with no desire to bias the further observations of Mr. Lapham, that I have used just such a one-fourth as he describes, but have ultimately discarded it, and in favor of a one-sixth marked by the maker 180°, which I find will do all the work (penetration included) of the one-fourth of 50° and a great deal more besides. But behind this one-sixth, I hold a one-tenth of 100° balsam angle in reserve, for work where " penetration" is ruled out.

As to the other point suggested. Is there not undue weight attached to this "penetration?" With me, I have often been obliged to take special measures to *get rid* of this function, and for this purpose use the Beck illuminator, which gives me only surface structures. I mention the fact, but am willing to admit that a certain amount of " penetration" is at times desirable, and should be provided for as perfectly as possible.

Mr. Lapham recommends a four-tenths of 100°. I have often thought that such a glass, or a half-inch of the same angle, would be desirable, and, as a *luxury*, am still of the opinion. It must be remembered that he very properly rules out the item of cost; while on the other hand I have made it a study to avoid expense, where my opinion has been solicited in the matter of selection of objectives.

Mr. Lapham recommends a certain number of low-angled glasses, all others to be of the highest angles, and of the best quality and well corrected. Why not amend this by insisting that all glasses have the highest attainable angles? selecting of such the one suitable for the work in hand.

Again, Mr. Lapham states that we have no need to consider objectives of lower power than the half-inch for " they are not made with angles sufficiently high to injure penetration." Here is involved an error in fact, one quite pardonable too. Not long since a German gentleman remarked to the writer that it seemed to him that when the London opticians *demonstrated* a certain optical law, some Yankee optician would be just mean enough to make an objective that would upset the whole arrangement! And it is perforce of this *fact* that Mr. Lapham becomes involved in his very pardonable error. Now I have in my possession a two-thirds of the Messrs. Spencers of 48° aperture. They have lately made to my order an inch of 47°. The clear diameter of both glasses is about the same; the working distance of the two-thirds is 25-100ths of an inch, while that of the inch is but 13-100ths of an inch, and the "penetration" of the latter (so-called) is less than that of the 2-3ds. In fact the penetration of the inch is "injured," as compared with one of 30°. The reader will notice that Mr. Wenham's pet theory gets also into grief. I assure Mr. Lapham that this new inch is a glorious glass, penetration or no penetration.

Finally, I desire to extend to Mr. Lapham personally my thanks for his instructive and interesting article. It will be a pleasure, and, I doubt not, profit to compare notes with him, and, in the words of my very generous opponent, Prof Hitchcock, I will add that all I desire is "the facts." Mr. Lapham, by the way, will be pretty sure to catch it, about that 180°, from Prof. II., on the "impossible" and the "absurd."

It may be well enough to add that my first impulse was to write Mr. Lapham privately for an interchange of personal experiences, but on reflection chose this, the more public plan, hoping to enlist the attention of others in the same direction.

OLEOMARGARINE AGAIN.

Ed. Am. Jour. Microscopy:—The article giving " The Microscopical Examinations of Oleomargarine," by Prof. Michels, which appeared in a recent issue of this journal, was perused by me with much interest, and I at once resolved to repeat the experiments as detailed.

The first thing requisite was to secure authentic specimens of oleomargarine. In order to do this, I solicited the friendly aid of a prominent gentleman of this city, who is also well known in business circles in New York. I also wrote, independently, to a gentleman in your city, urging him to obtain for me the samples required.

Desiring also to procure a specimen of genuine dairy butter, I applied to a well-known citizen of this city, who procures his supplies directly from a farmer.

After an interval of ten days, I was, in response to my solicitations, in possession of three samples of oleomargarine, and one sample of pure dairy butter. The three samples of oleomargarine are directly from the manufacturers at New York city, and will be referred to as Nos. 1, 2, and 3.

The *four* samples were subjected to examination under the microscope ; the objectives used were a 1-4th inch of 100° by Tolles, and Spencers late duplex 1-4th of 180°.

Samples Nos. 1, 2 and 3, and also the specimen of pure dairy butter, showed many crystals of chloride of sodium ; the crystals furnished by the pure butter were, however, cleaner, and more acceptable generally than those exhibited by Nos. 1, 2 and 3.

Samples Nos. 1, 2 and 3, in addition to crystals of chloride of sodium, displayed other crystals, those of nitrate of soda being prominent ; while forms closely resembling crystals of *cholesterine* were found in considerable numbers.

Regarding Nos. 1, 2 and 3, it may be remarked that Nos. 1 and 2 were, to outward appearance, tolerably fair imitations of genuine butter, and might by the ordinary purchaser be accepted and bought as such. No. 3 was a poorer counterfeit, and would probably be rejected by most buyers ; but if mixed in equal parts with the genuine article, the mixture would be likely to deceive the purchaser.

In samples Nos. 1 and 2 the microscope displayed the " feathery crystals" (margarine) described by Prof. Michels, although these were not constantly present in every field examined. By moving the slide, other fields were brought to view in which these crystals were much more prominent than those given in the cut accompanying Prof. Michels' paper. The crys-

26 Microscopy.

tals of nitrate of soda and cholesterine (?) were to be seen in every slide prepared from the samples named.

Besides the crystals named, samples Nos. 1 and 2 gave " suspicious cells " in large numbers, accompanied by shreds and tissue fibres, many in a broken down condition, while others seemed to be in a tolerably normal state, sufficient almost to establish the presence of voluntary muscle. Bundles of these fibres were closely examined with the duplex 1-4th, under powres as high as 1,600 diameters, with the endeavor to bring out the transverse markings; this was not accomplished—owing, doubtless, to the nature of the *vehicle* (my observations in this direction will be continued). Many of the bundles seen were completely broken down, and the elementary fibres detached.

Sample No. 3, when examined under the microscope, displayed fewer of the "feathery crystals;" nevertheless, this was the most "suspicious" specimen of the three; there were multitudes of "suspicious cells," shreds and patches of tissue, in a more or less broken down state. In one field I felt tolerably sure of finding encysted hydatids. This slide unfortunately, was accidentally destroyed by the water getting under the cover.

To give an intelligent description of this material (No. 3) would require far more time to its study than I have at my command at present; but I hope to attack it again, and at an early day. Suffice it now to say that this specimen contains *very many* "suspicious" elements, and that its behavior under the objective (here I also include Nos. 1 and 2), is—with the exception of the crystals of chloride of sodium, and the presence of a few fatty globules—totally different from that of pure dairy butter.

The sample of pure dairy butter gave fields just as represented by Prof. Michels, with the exception that crystals of the chloride were almost constantly present.

Having established, to my own satisfaction, at least, the integrity of the observations of Prof. Michels, I therefore hold that the gentleman should be regarded in the light of a public benefactor; the matter he has presented will be found worth serious and careful investigation.

In one of my recent lectures before the medical class of this

College, on the *Entozoa*, I quoted (substantially) from Dr. Roberts, as follows:

"A marvelous light has been thrown in recent years on the zoological position of the Entozoa, chiefly by the researches of *Siebold* and *Van Beneden*. It has been ascertained that the hydatid worm found in man, constitutes the *encysted* phase in the developement of a very minute tape-worm which infests the dog. The tape-worm in question is the *Tænia echinococcus*; the entire adult animal is so small that it scarcely exceeds the size of a millet seed. It consists of but three segments, of which three, the last only is fruitful. When this segment arrives at maturity, it is cast off, and a new one developed in its place. Myriads of these worms are sometimes found in the intestine of the dog, and their eggs are discharged in countless numbers with the excrements, the eggs so discharged are scattered far and wide; and some of them find their way with the food into the stomachs of men and other creatures suitable for their development. Arrived there, the embryo is liberated; and after penetrating the mucus membrane, it burrows its way, or is carried by the blood current to some distant organ, where it is arrested. Having thus lodged itself, it presently reappears as a hydatid vesicle, in which are developed the echinococci, as before explained. Dogs in their turn become infested with the corresponding tænia by feeding on the offal of slaughtered sheep, pigs, etc., which had been infested with hydatids. The echinococci therein contained develop in their intestines into the tænia echinococci, and thus the circle of transformation and development recommences."

By similar cycles of transformation and development, do we arrive at a class of parasites known in medicine as the ectozoa. This term may be said to include, or be applied to, worms or larvæ of insects that have been introduced into the intestinal canal by accident. Animalcules, such as the hair worm, grub of the fly, may be mentioned; also the larva of the bee, the spider, etc. Among animals, the disease known popularly as the botts, to which the horse is frequently a victim, is caused by such animals swallowing the ova of the *oestrus* or gad-fly.

That oleomargarine manufactured from refuse animal fats,

in the manner described by Prof. Michels, and at a temperature not above 120° F., may be a highway through which " eggs so discharged " find their way into the stomachs of men, is too palpably evident to need further comment.

INDEX.

A

Acetate of soda, solution, 329.
Acme microscope, 84.
Adjustable objectives, 148.
Adjustable glasses, 129.
American stands, 17.
American stands compared with German, 18.
American histological stand, 39.
Analysis, chemicals for, 322.
Analysis, chloride of sodium, 328.
Analysis for albumen, 337.
Analysis for glasses, 103.
Analysis, earthy phosphates, 331
Analysis, phosphoric acid, 330.
Analysis, sugar, 332.
Analysis, sulphuric acid, 331.
Analysis, urea, 323, 325.
Analysis for sugar, 335.
Angular aperture, what is it? 93.
Angular aperture defined, 93, 370.
Angular aperture illustrated, 94.
Angular aperture, how to measure, 95.
Angular aperture, versus working distance, 115, 121.
Angular aperture and central illumination, 361.
Angular aperture and penetration, 118.
Angle high glasses, 103.
Apparatus for micro-chemical use, 320.
Apertures, angular, 93,

B

Aperture, angular balsam, 123.
Artificial light, 187.

Balsam apertures, 123.
Balsam angles, advantages of, 141.
Balsam angle, management of 237,
Balsam angle, and working distance, 142.
Balsam angle, low objective 144.
Baryta solution, 324.
Bausch & Lomb's microscopes, 50.
Bausch & Lomb's microscope, professional, 53.
Bausch & Lomb's microscope, students, 56.
Bausch & Lomb's microscope, stands, 344.
Bausch & Lomb's objectives, 346.
Beck's microscopes, 72.
Beck's microscope stands, 347.
Beck's vertical illuminator, 217.
Beck's vertical illuminator, how to use, 221
Binocular, objections to, 30.
Binocular, versus monocular, 29.
Biological microscope, 68.
Bull's eye condenser, 225.
Bull's eye condenser, how to use, 227.
Bullock's microscope, first-class, "A. 1." 59.

405

Bullock's microscope, 59.
Bullock's microscope, D stand, 57.
Bullock's microscope, small, best, 67.
Bullock's microscope stands, 343.

C

Carbonate of soda, 324.
Carpenter on object glasses, 97.
Centennial stand, Zentmayers, 34.
Centennial stand and A. 1. compared, 63.
Central light and high angles, 257
Choice of objectives, 397.
Choice of objectives for regular work, 202,
Chemicals for analysis, 322.
Chloride of sodium, 327.
Chromate of Potash, 327.
Collar adjustment, objections to, 208.
Concentric rotating stage, 25.
Condenser, Bull's eye, 225.
Condenser, small diameter, 308
Contributions to the Medical News, 358.
Covering glass guage, 216,
Covering glass, selection of, 213.

D

Daylight, how to use it, 183.
Diatoms, examinations of, 255.
Diatoms for test, 251.
Diatoms, order for test, 254.
Diatoms resolution as objective test, 249.
Diatoms resolution, with high powers, 245.
Diatoms resolution, with low powers, 245.

Diatoms, resolution, 243.
Draw tube, advantages of, 136.
Dry mounts, list of, 305.
Dry mounts, with high objectives, 298.
Duplex glass, the first, 9.

E

Essentials of a reliable stand, 21.
Examination of morbid products, 206.
Examination of urinary deposits, 203.
Examination of oleomargarine, 400.
Eye pieces, 256.
Eye pieces high, 146.
Eye pieces fittings, 33.
Eye pieces solid, 156.
Eye pieces should fit loosely 204.
Eye training, 237.
Eye training illustrated, 239.

F

Family microscope, 57.
Fasoldt's micrometer, 357.
Fehling's solution, 333.
Ferro cyanide of potassium solution, 329.
Field, flatness of, 126.
Fighting objectives, 117. 140.
Flatness of field, 126.
Four system glasses, 9.

G

Glasses, high angle, 103.
Glasses, high angle history of 100,
Glasses, testing aperture, 192.
German stands compared with American, 18.

German student lamp, 187.

H

Handling objectives, 259.
High angles, 103.
High angles balsam, 125.
High angles, discursion of, 359.
High angles, objectives, 100.
High, eye piecing. 146.
High objectives, with dry mounts, 298.
Higher powers, work with. 235.
History of high angle objectives, 100.
Histological stand, 39.

I

Illumination, artificial, 187.
Illumination, daylight, 183.
Illumination small best, 189.
Illumination, sunlight. 186.
Illumination, varieties of, 183.
Illuminator, Beck's vertical, 217
Illuminator, modified, 219.
Illuminator, Wenham's reflex, 157.
Illuminator, Wenham's with sunlight, 191.
Illuminator, Woodward's, 163.
Immersion, 125.

K

Kerosene oil immersion lens, 316.

L

Lens, Tolles' traverse, 179.
Lessons on seeing with the microscope, 261.
Lesson first, 261.

Lesson second, 262.
Lesson third, 265.
Lesson fourth, 266.
Lesson fifth, 268.
Lesson sixth, 270.
Lesson seventh, 271.
Lesson eighth, 273.
Lesson ninth, 274.
Lesson tenth, 280.
Lesson eleventh, 282.
Lesson twelfth, 284.
Low balsam angles, 144.
Low powers, working with, 225.

M

Management of high angles, 257.
Manipulations, microscopic, 157.
Marais' approximate tubes, 330.
Measuring bottle, 324.
Medium power of glass 144.
Micrometer, necessity for, 252.
Micrometer, Fasoldt, 357.
Micrometer, Rogers', 356.
Micro-chemical apparatus, 320.
Micro-chemical examination of urine, 318.
Microscope, Acme, 84.
Microscope, Bausch & Lomb, 51.
Microscope, Beck's, 72.
Microscope, Bullock's, 59.
Microscope. dissecting, Beck's 79.
Microscope, "D," 67.
Microscope, economic. 72.
Microscope, family, 57.
Microscope, first-class A 1, 59.
Microscope, international, 348.
Microscope, investigator, 345.
Microscope, new biological, 68.
Microscope, new histological, 79.
Microscope, monocular vs. binocular, 80.

Microscope, national, 29.
Microscope, popular, 72.
Microscope, professional, 53.
Microscope students, 56.
Microscope, Sidle's, 84, 351.
Microscope selection, 147.
Microscope, Tolles , 47.
Microscope, Zentmayer, 34.
Microscope, use and abuse of 375.
Microscopic manipulations, 157
Mirrors, concave, 32.
Mirrors, hangings strong, 38.
Mirrors mountings, 32.
Mirrors, plain, 31.
Mirrors used as condensers, 229.
Mirrors where attached, 226.
Monocular vs. binocular, 29, 80.
Moller's test-plates, measurements of, 253.
Mounting of objectives, 129.

N

Names of microscopic dealers, 341.
Nobert's test-plate, 254.

O

Objections to mechanical stages 24.
Objections to binoculars, 30.
Objectives, adjustable, 148.
Objectives, broad guage, 230.
Objectives, balsam, 141.
Objectives, choice of, 397.
Objectives, Dr. Carpenter on 97.
Objectives, fighting, 117, 140.
Objectives, high angle, 97.
Objectives, high angle, testing of, 107.
Objectives, low balsam angle, 144.

Objectives, one-sixth preferred to one-fiftieth, 109.
Objectives, immersion, 125.
Objectives, experience with, 202.
Objectives, for physicians, 207.
Objectives for regular work, 202.
Objectives, test with diatoms, 249.
Objectives, testing aperture of 192.
Objectives, testing, 198.
Objectives, wide angle, what is it ? 137.
Objectives, working distance of, 108.
Objectives, mounting, 129.
Objectives, oil immersion, 310.
Objectives, Spencer's, 353.
Objectives, Tolles', 354.
Objectives, performance of, 368.
Observer's position, 246, 289.
Oil immersion objectives, 309.
Oil immersion, objectives, experience with, 315.
Oil immersion objectives of Zeiss, 310.
Oleomargarine examinations, 400.

P

Penetration and angular aperture, 118.
Physicians, objectives best for, 207.
Physics, elementary hints on, 393.
Podura, scale resolution of, 299.
Position of observer, 246, 289.
Professional microscope, 53.

R

Rising of the object, 276.

Reaction of urine with tests, 339.
Rogers' micrometers, 356.

S

Selection of covering glass, 213.
Selecting a stand,
Sliding tube, advantages of, 44.
Short tube stands, 25.
Small light best, 189.
Small kerosene lamp, 189.
Society screw, 232.
Society screw not sufficient for high angles, 88.
Specific gravity and temperature, 340.
Spencer's objectives, 353.
Spencer's one inch of 50°, 230.
Stage, thin, improvised, 39.
Stage swing and mirror best, 91.
Stage, concentric, rotating, 25.
Stage, mechanical objections to 24.
Stage, room necessary, 26.
Stage used by the author, 27.
Stand, Acme, 84.
Stand, American, 17.
Stand, American Centennial, 34.
Stand, Bausch & Lomb's, 50.
Stand, Beck's, 72.
Stand, Bullock's, 59.
Stand, essential of a reliable, 21.
Stand, Economic, 72.
Stand, Histological, 79.
Stand, " D," 67.
Stand, Investigator, 345.
Stand, International, 348.
Stand, large and small, 26.
Stand, large " BB," 47,
Stand, large " A," 48.
Stand, new biological, 68.
Stand, new National, 80.
Stand, first class " A. 1." 59.

Stand, small best, 67.
Stand, popular, 72.
Stand, Sidle's, 84.
Stand, Tolles', 47.
Stand, students', 49, 56.
Stand, small and cheap, best, 92.
Stands, with short tubes, 25.
Stands, Zentmayer's, 34.
Students microscopes, 49, 56.
Sunlight illumination, 186.

T.

Table for work, 220.
Temperature and Urine specific gravity, 340.
Testing high angles, objective, 107.
Thin stage necessary, 90.
Tight fitting eye pieces, 204.
Tolles' first duplex, 9.
Tolle's microscopes and objectives, 354.
Tolles' traverse lens, 179.
Tolles' large " BB " stand, 47.
Tolles' large " A " stand, 48.
Tolles' students' stands, 49.
Traverse lens, 179.
Tube casts, no genuine seen, 319.
Tube sliding, advantages of, 44.

U.

Urinary constituents, proportion of, 339.
Urinary deposits, examinations of, 203.
Use and abuse of the microscope 375.

V.

Volumetric analysis, 318.

W.

Weights recognised, 321.
Wenham's reflex illuminator, 157.
Wide apertures, lessons in the use of, 261
Wide apertures, management of 258.
Wide angle objectives, 137.
Wide angle objectives and collar adjustment, 139.
Wide angle objectives and illumination, 140.
Wide angle objectives and working distance, 137.
Work table, 229.
Working with higher powers, 235.
Working with lower powers, 225.

Working distance of high angles, 137.
Working distance of high balsam angles, 143.
Working distance of objectives, 108.
Working distance vs. angle aperture, 115, 121.
Woodward's illuminator, 163.
Woodward's illuminator modified, experience with, 177.

Z.

Zentmayer's stands, 34, 355.
Zentmayer's American Centennial, 34.
Zentmayer's American Histological, 39.

SPECTACELS;

AND HOW TO CHOOSE THEM.

BY C. H. VILAS, A. M., M. D.

Professor of Diseases of the Eye and Ear in the Hahnemann Medical
College, and Ophthalmic and Aural Surgeon to
the Hahnemann Hospital,
Chicago, etc., etc.

This book is full of interest and instruction to readers of all
classes. Designed more especially for the profession, it nevertheless does not deal in technicalities or obscure terms and
will doubtless find a large demand among non-professional
people.

In a clear and comprehensive manner it treats of a most
important subject, a subject concerning which, far too many er-
roneous and dangerous notions prevail. It's aim is to dissemin-
ate knowledge, and to prevent the common haphazard custom
of choosing Spectacles, a custom so often disastrous to vision or
fruitful to discomfort. The Author points out how, within a
comparatively recent period, what was once veiled in mystery,
has attained a sure place among the sciences ; how the knowledge of the proper construction, adaptation, and uses of Spectacles has grown to great proportions, and has led to a revolution in the treatment of eye affections. Many popular delusions
will be dispelled by reading this book. It is valuable alike to
the physician and the layman. The former may utilize it in
practicing his profession ; the latter will prepare himself to
avoid errors fatal to vision.

All the new and useful cases of trial lenses are described and
illustrated and their respective merits and demerits are pointed
out. It is clearly shown how the low prices, at which a good
set of trial lenses can be obtained, will enable the physicians in
general practice to give attention to the fitting of Spectacles
and to treating ordinary defects of the vision, thus preventing
the frauds so often practiced by itinerant venders.

The Author's well known accomplishments in this department of science, have eminently prepared him to skillfully
treat so important a subject. The work is illustrated and full
of practical hints.

Price bound in cloth, $1.00.

DUNCAN BROS., Publishers.

131 & 133 S. Clark St.,   *CHICAGO.*

HOW TO BE PLUMP.

OR

TALKS ON PHYSIOLOGICAL FEEDING.

BY

T. C. DUNCAN, M.D.,

*Author of a Professional Treatise on Diseases of Infants
and Children and their Homœopathic Treat-
ment; Editor " United States Medical
Investigator, etc. etc."*

Finely bound in Cloth - - - - **Price, 50 Cents.**
It is an admirable work on hygiene, as far as it relates to eating—*Ohio. Farmer.*

These " Talks " have their own charm, for most of us like to be plump. —*Women's Journal, Boston.*

Certainly all *thin* people should ponder well its suggestions and put them into practice.—*Voice of Masonry.*

A *brochure* which many persons of lean inclinations will be glad to read and may be able to profit by.—*Banner of Light.*

The little volume contains much sensible professional advice upon healthy and morbid digestion, and the occasions of it.—*Zion's Herald.*

The book is intended to indicate how anyone may become fat, fair and jolly. and a careful perusal affords proof that the author is fully conversant with his subject, and that he has done it full justice.—*The Chicago Grocer.*

It is a common sense volume, that believes in rational and practical methods of preserving health, beauty and happiness. It is not a receipt book of impossibilities; it is a bright, genial book, that understands itself from first to last.—*The Chicago Cosmopolitan.*

We do most warmly applaud its purpose, and especially commend its philosophy to the unnaturally lean, and still more to mothers in the training of delicate children whose physical stamina hardly equals that of a full-grown rye-stalk. Theirs is a leanness to be built up into steady strength and picturesque plumpness of limb and face.—*The Standard.*

CHICAGO, Sept 18, 1878.
The case referred to on page 45 of " How to be Plump" is our little boy, who was certainly rescued from death's door. This summer he began to run down again. After Dr. Duncan had tried several medicines without benefit he again advised " inunction," and again with the same happy effect. The little fellow is now plump and well. CHARLES WALES.

☞May e ordered through your physician, or newsdealer, or will be sent direct on receipt of price.

DUNCAN BROS., Publishers,
113 *Madison and 131 & 133 Clark St., Chicago.*